"Story-telling helps peo   d,
from the campfire to cl;   ns
volume will aid every(nd
phrases that people use ᴗ _   of
energy demand."

— Jim Skea, Professor of Sustainable Energy,
Imperial College London, UK

"This book is essential reading for policy makers who think
they can solve climate change by promoting energy efficiency
and adding more renewable energy sources. *Energy Fables* shows
that energy demand should be the main topic of research and
policy intervention."

— Kris De Decker, Low-tech Magazine, Spain

"*Energy Fables* provides a challenge to the way we think about
energy. Some contributions will doubtless be controversial, but
the sustained focus on how energy is used and the social construc-
tion of energy demand is thought-provoking and welcome."

— Nick Eyre, Director of the Centre for Research
into Energy Demand Solutions, UK

ENERGY FABLES

Energy Fables: Challenging Ideas in the Energy Sector takes a fresh look at key terms and concepts around which energy research and policy are organised.

Drawing on recent research in energy and transport studies, and combining this with concepts from sociology, economics, social theory and technology studies, the chapters in this collection review and challenge different aspects of received wisdom. Brief but critical introductions to classic notions like those of 'energy efficiency', 'elasticity', 'energy services' and the 'energy trilemma', together with discussions and analyses of well-worn phrases about 'low hanging fruit' and 'keeping the lights on', articulate aspects of the energy debate that are often taken for granted. In re-working these established themes and adding twists to familiar tales, the authors develop a repertoire of new ideas about the fundamentals of energy demand and carbon reduction.

This book presents a valuable and thought-provoking resource for students, researchers and policy-makers interested in energy demand, politics and policy.

Jenny Rinkinen is a Researcher in the Centre for Consumer Society Research at the University of Helsinki, Finland.

Elizabeth Shove is a Professor in the Department of Sociology, Lancaster University, UK, and was PI of the DEMAND Research Center. She is the author/editor of numerous titles, including *The Nexus of Practices:*

Connections, Constellations, Practitioners (Routledge, 2017, with Allison Hui and Theodore Schatzki) and *Sustainable Practices: Social Theory and Climate Change* (Routledge, 2013, with Nicola Spurling).

Jacopo Torriti is Professor of Energy Economics and Policy in the School of the Built Environment, University of Reading, UK, and author of *Peak Energy Demand and Demand Side Response* (Routledge, 2016).

ENERGY FABLES

Challenging Ideas in the Energy Sector

Edited by Jenny Rinkinen, Elizabeth Shove and Jacopo Torriti

Routledge
Taylor & Francis Group
LONDON AND NEW YORK

from Routledge

First published 2019
by Routledge
2 Park Square, Milton Park, Abingdon, Oxon OX14 4RN

and by Routledge
52 Vanderbilt Avenue, New York, NY 10017

Routledge is an imprint of the Taylor & Francis Group, an informa business

© 2019 selection and editorial matter, Jenny Rinkinen, Elizabeth Shove and Jacopo Torriti; individual chapters, the contributors

The right of Jenny Rinkinen, Elizabeth Shove and Jacopo Torriti to be identified as the authors of the editorial material, and of the authors for their individual chapters, has been asserted in accordance with sections 77 and 78 of the Copyright, Designs and Patents Act 1988.

British Library Cataloguing-in-Publication Data
A catalogue record for this book is available from the British Library

Library of Congress Cataloging-in-Publication Data
Names: Rinkinen, Jenny, editor. | Shove, Elizabeth, 1959- editor. | Torriti,
    Jacopo, editor.
Title: Energy fables : challenging ideas in the energy sector / edited by
    Jenny Rinkinen, Elizabeth Shove and Jacopo Torriti.
Description: Abingdon, Oxon ; New York, NY : Routledge, 2019. |
    Includes bibliographical references and index. |
Identifiers: LCCN 2018059847 (print) | LCCN 2019002363 (ebook) |
    ISBN 9780429397813 (Master) | ISBN 9780367027759 (hardback :
    alk. paper) | ISBN 9780367027797 (pbk. : alk. paper) | ISBN
    9780429397813 (ebook : alk. paper)
Subjects: LCSH: Energy resources development. | Energy consumption. |
    Energy policy. | Energy industries.
Classification: LCC HD9502.A2 (ebook) | LCC HD9502.A2 E543847
    2019 (print) | DDC 333.79—dc23
LC record available at https://lccn.loc.gov/2018059847

ISBN: 978-0-367-02775-9 (hbk)
ISBN: 978-0-367-02779-7 (pbk)
ISBN: 978-0-429-39781-3 (ebk)

Typeset in Bembo
by Swales & Willis Ltd, Exeter, Devon, UK

Printed and bound in Great Britain by
TJ International Ltd, Padstow, Cornwall

CONTENTS

CONTRIBUTORS

Noel Cass is a Researcher in the Department of Organisation, Work and Technology at Lancaster University. He has worked on research projects dealing with a range of environmental and energy issues. Recent publications include 'Infrastructures, intersections and societal transformations', with Tim Schwanen and Elizabeth Shove, in *Technological Forecasting and Social Change* (2018).

Mike Hazas is Reader in the School of Computing and Communications at Lancaster University. Mike works at the interface of human-computer interaction and studies of social practice, striving for nuanced understandings of everyday life and sustainability through both qualitative and quantitative data. He co-directs the socio-digital sustainability research team at Lancaster.

Greg Marsden is Professor of Transport Governance at the Institute for Transport Studies at the University of Leeds. Current work examines the transition to a low carbon transport system and how 'smart mobility' will impact transport governance. He is chair of the Commission on Travel Demand, looking at how to deal with alternative demand futures. He has served as an adviser to the House of Commons Transport Select Committee and regularly advises national and international governments.

Janine Morley is a Senior Research Associate in the DEMAND Centre at Lancaster University. She researches the relationships between everyday practices, technologies and consumption, and is interested in questions raised by climate change and environmental sustainability. Her work applies and develops theories of practice, and has most recently focused on digital technologies, their evolving roles in everyday life and the implications for energy demand.

Jenny Rinkinen is a Researcher in the Centre for Consumer Society Research at the University of Helsinki, Finland. She is a social scientist specialising in energy demand, sustainable consumption and resource use. Her recent research has focused in particular on changes in energy-intensive practices such as housing, heating, cooling and food supply.

Sarah Royston is a Research Fellow at the University of Sussex, and is part of the DEMAND Centre. Sarah works on the Invisible Energy Policy project, and was previously employed at the Association for the Conservation of Energy and at Keele University.

Jan Selby is Professor of International Relations at the University of Sussex. His research focuses on political ecology and environmental security, and especially on the intersection of issues to do with water, climate and energy. He leads the Invisible Energy Policy project within the DEMAND Centre.

Elizabeth Shove is Professor of Sociology at Lancaster University, and was PI of the DEMAND Centre. She has written about social practices, everyday life and energy demand, and has recently co-edited *Infrastructures in Practice: the Dynamics of Demand in Networked Societies* with Frank Trentmann (Routledge, 2018).

Yolande Strengers is an Associate Professor of Sustainability and Urban Planning at the RMIT University. Yolande is a sociologist of technology and design and co-leader of the Beyond Behaviour Change research programme at RMIT. She researches the changing role of digital and smart technologies in everyday life and their implications for energy and sustainability outcomes.

Jacopo Torriti is Professor of Energy Economics and Policy in the School of Built Environment, University of Reading and a Co-Director of the Centre for Research into Energy Demand Solutions (CREDS). His research focuses on various aspects of flexibility in energy demand, including the timing of people's activities, peaks, policies, pricing, technologies and economic incentives. He is author of *Peak Energy Demand and Demand Side Response* (Routledge, 2016).

Gordon Walker is Professor in the Lancaster Environment Centre and recently Co-Director of the DEMAND Centre. He is a human geographer interested in the relations between technology, environment, justice and practice. He has researched across a wide range of cases and concerns – including energy demand, community energy, fuel poverty, thermal risks, zero carbon homes, flooding and air quality – drawing on concepts from human geography, science and technology studies, and normative theory.

ACKNOWLEDGEMENTS

This work was supported by the DEMAND: Dynamics of Energy, Mobility and Demand Research Centre funded by the Engineering and Physical Sciences Research Council (grant number EP/K011723/1) as part of the Research Council's UK Energy Programme and by EDF as part of the R&D ECLEER Programme. For more on the DEMAND Centre see www.demand.ac.uk.

1

INTRODUCTION

Elizabeth Shove, Jacopo Torriti and Jenny Rinkinen

Energy research and energy-related policy making are informed by terms, ideas and stories that reproduce certain ways of thinking about problems and responses. As in other fields, phrases enter common usage, concepts become taken for granted, and shared vocabularies form. Disciplines and approaches build on these foundations, often forgetting that what seems like obvious, or common wisdom has a history: it is not set in stone nor is it uncontested or uncontroversial. The essays included in this collection explore different aspects of what we refer to as 'fables' in the energy world.

The phrases and sayings about which we write are not simply 'made up', or fictional in the sense of being products of the imagination alone. Rather, our point is that they act as orienting narratives; as tales that have conceptual foundations that have become invisible, worn smooth through use and submerged within familiar discourses in government, as well as in research and teaching.

In reviewing and revisiting a selection of terms that have this fable-like status we have two main goals. One is to introduce and also problematise the concepts we discuss – reminding old hands and newcomers alike that recurrent refrains (such as, 'first pick the low hanging fruit'), imperatives (to keep the lights on) and policy injunctions (engage with the energy trilemma) reproduce ideas that need not, and perhaps should not, be taken at face value. The second is to enrich the repertoire of concepts circulating in this field and to create opportunities for disciplines and approaches outside the realm of energy research to make useful contributions. On both counts, new 'stories' and vocabularies are needed.

The idea for this collection came from this double realisation.

Although approaches differ, the essays included here have certain features in common. Each sets out distinctive characteristics of the 'fable' or topic in question and each offers a critical analysis of the ideas on which the phrase or statement depends and of the challenges and issues that follow. The result is more than a glossary or index of key terms. In detailing what lies behind dominant discourses and sayings, contributors articulate and critique assumptions and conventions that have become embedded in the energy field. We do not push the analogy with Aesop's fables very far, but each chapter begins with an emblematic image and a short statement, making a link to this tradition. This is partly for fun, but also as a reminder of the power of metaphor and discourse.

Terms and languages are always changing, and we do not pretend to offer a final, rival or definitive lexicon. Instead, our purpose is to make the familiar strange again in order to generate and promote critical reflection about some of the core ideas around which energy and transport studies revolve. In detail, the collection examines eleven terms: energy demand; energy services; efficiency; rebound; elasticity; picking low hanging fruit; keeping the lights on; promoting smart homes; the energy trilemma; flexibility; and non-energy policy. This is not an exhaustive list, but in combination, these topics represent and exemplify dominant approaches in energy and transport policy and research.

Each 'fable' can be read as a stand-alone contribution and each is designed to inform and inspire students, teachers, policy makers or researchers with an interest in that topic. Reading them together gives a stronger sense of how related traditions and schools of thought 'hang together' and support each other. In designing the collection, we have sought to make some of these connections plain.

The first two chapters following this introduction (on demand and on services) show how energy is conceptualised as something that is abstracted from what people do, and from the histories, cultures and contexts in which energy demand is constituted.

The chapters in Part II (efficiency, rebound and elasticity) explore a series of linked ideas about the character of provision and consumption. All three terms are rooted in engineering and/or economics and all share a tendency to treat energy as a topic in its own right (as a resource, a component of an input-output equation or as a standardised commodity). In brief, notions of efficiency are in essence about delivering *the same* service but with fewer units of energy. The terms 'rebound effect' and 'elasticity' both emphasise the role of price (per unit of energy) in determining changes in demand. For example, the concept of rebound relates to the idea that the savings associated with efficiency gains might trigger increases in other (often non-energy) carbon intensive commodities and services. Meanwhile, the idea of elasticity concerns the anticipated impact of an increase or decrease in price on demand. In discussing these concepts we remind readers of the work involved in constituting 'energy' as a distinctive field of research and intervention.

The contributions in Part III focus on specific injunctions: 'first pick the low hanging fruit', 'keep the lights on' and 'promote smart homes'. In all these cases, energy (as used in buildings, or for transport) is viewed as a quantifiable resource used in delivering pre-defined services. Picking 'low hanging fruit' refers to opportunities to 'harvest' quick wins in terms of energy efficiency, and to do so without compromising the level of service provided. Assumptions about the non-negotiability and apparently 'fixed' status of energy services – and of the societal need for them – are also reproduced in the rhetoric of 'keeping the lights on'. In revisiting this idea we ask which lights need to be kept on, and why is lighting in any case so important? The third example, which has to do with the potential for using 'smart' technology and controls to manage domestic energy demand, depends on a similarly powerful set of ideas about present and future ways of life, and about the character of 'consumer' response.

The three chapters in Part IV deal with themes relating to policy. The first of these concerns the energy trilemma. The notion of an energy trilemma suggests that the goals of promoting energy security, affordability and decarbonisation are all important, but also in tension. This is so in that strategies that focus on one corner of the trilemma – like decarbonising supply – might make it harder to achieve other goals, such as providing 'affordable' power. Although talk of the trilemma is pervasive, it routinely sidesteps questions about the scale, the extent and the character of demand.

By contrast, demand seems to be crucial for those interested in flexibility and demand side management. Ideas about demand are also fundamental for debates about policies and interventions that have a bearing on the reduction and the escalation of energy consumption. So how is demand conceptualised?

Until recently, detailed questions about when and where energy is used were of relatively little concern partly because solid fuel, gas, and oil can be distributed and stored. An influx of more renewable sources of supply complicates this approach – bringing with it a new agenda about exactly when and where different forms of energy (and especially electricity) are used, and about the scope for shifting demand in space and time. From this point of view, the concept of 'flexibility' draws attention to the fact that both the location and the timing of energy consumption is indirectly shaped by policies that are not specifically concerned with energy or carbon emissions as such, but that have profound consequences for both.

In elaborating on the implications of this observation, the last fable in the collection takes issue with the view that the policies that matter for energy or transport are those that are explicitly or primarily concerned with energy or with transport.

Energy Fables highlights threads and lines of reasoning that run through the energy landscape. The list of terms and the topics we discuss is not exhaustive and others – for example, fuel poverty, comfort or storage – could be explored in similar terms.[1] In the postscript we take stock of the practical and policy implications of reviewing and sometimes overhauling the concepts and assumptions we discuss. We also recognise the everyday politics of energy and transport research: it is no accident that problems are defined and framed as they are, or that so much is invested in only some lines of enquiry and investigation. Dominant paradigms and interpretations of practical value and policy relevance interlock, and do so in ways that perpetuate many of the fables discussed here. While it would be naïve to expect new discourses and debates to spring up overnight, even small shifts of perspective generate new problem definitions, and with them new opportunities for intervention. Which new fables emerge, and how concepts of energy change and develop in the future remains to be seen. The fact that this book has no final punchline, and no one unifying conclusion, is not an accident or an oversight: it is because the field is open and because our purpose is to provoke and fuel future discussion, not to bring it to an end.

Note

1 See the online 'DEMAND dictionary of phrase and fable' at www.demand.ac.uk/demand-dictionary for more examples.

PART I

What is energy for?

PART I

What is energy for?

2

ENERGY DEMAND

Jenny Rinkinen and Elizabeth Shove

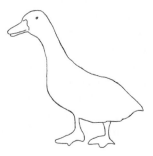

*Some creatures are more insistent than others,
and quacking ducks can be especially demanding.
This chapter explores the social foundations of
energy demand.*

Introduction

Energy demand, usually indicated by the amount of energy consumed in different sectors, is a key factor in national and international energy and climate change policy. Since carbon emissions are significantly associated with the production and distribution of energy, demand reduction is a priority. This is evident in a recent strategy produced by the UK's six 'End use energy demand' research centres, which claims that 'reducing energy demand and improving energy efficiency provide the most promising, fastest, cheapest and safest means to mitigate climate change' (EUED, 2016, p. 5).

But what is energy demand, really?

The *Oxford English Dictionary* defines demand as 'an insistent and peremptory request, made of as right' (Oxford Dictionaries, 2018). In the energy world, demand has different meanings and associations: it often figures as the logical partner to 'supply'; it is tied to interpretations of 'need'; and it is the subject of various forms of intervention, including methods of 'demand side' management. As detailed below, all three of these interpretations treat energy demand as a social and economic phenomenon that exists in its own right.

This view is at odds with more sociological and more historical analyses of what energy is for and of how patterns of demand, in a more fundamental sense, emerge and change. As outlined by Shove and Walker (2014), such approaches suggest that it is not, strictly speaking, *energy* that is demanded. Rather, demand is for the services that energy makes possible: that is, for heating, lighting or mobility, for functioning televisions and laptops, for frozen food or for hot dinners. The demand for such services is, in turn, inseparable from the ongoing dynamics of social practice. From this point of view, it makes no sense to treat 'demand' as if it was in some way detached from the social arrangements and from the technologies and infrastructures of which societies are composed, and of which 'needs' are made.

In this chapter we distinguish between (a) definitions that interpret demand as the energy required to meet current and future needs and (b) those that take energy demand to be inseparable from the constitution of energy services and social practices. The first interpretation supposes that energy suppliers and policy makers are involved in meeting needs that exist, ready-made. In these accounts, demand (in the sense of what energy is for) is treated as given, rather than as a topic of research or policy intervention in its own right. By contrast, the latter position highlights the ongoing constitution of demand and the part that policy makers and others play in establishing and changing what people do and the energy requirements that follow.

Meeting demand

Engineers and economists generally think of energy as a 'resource' that is produced, distributed, and supplied in response to consumer demand. Energy is consequently treated as a standardised commodity, measured in agreed units (therms, kWh, Mtoe, etc.), and subject to various laws of the market. For example, if prices rise, consumers are expected to cut back on the amount of energy they use, and to consume (somewhat) more if costs go down. This relationship is complicated by different forms of 'price elasticity'. In practice, consumers are more willing or able to reduce some forms of energy demand than others. But in general, and alongside this uneven sensitivity to price, the consensus is that consumers buy the amount of energy they need unless there is some kind of shortage, and unless they cannot afford to do so. As a result, need and consumption are regularly taken to be the same. This line of argument is consistent with the view that producers and distributors aim to meet consumers' needs and design and organise systems of provision so that supply matches demand.

The scale of energy supply, now, and in the future, consequently depends on estimates of present and future demand. Established methods of demand forecasting involve judgements about trends in urbanisation, lifestyle, population and economic growth (Asif and Muneer, 2007). Not all forecasts anticipate increases in consumption, but those that do justify investment in more extensive systems of provision (more power stations, more roads etc.). In treating demand as an outcome of a host of economic and other consumer-oriented factors, forecasters treat the estimate of future supply as something that is quite unrelated to past and present technologies and infrastructures of provision. Put differently, they do not imagine scenarios in which demand grows *in response* to supply, nor do they consider the possibility that forecasts are themselves part of engendering future needs.

This is consistent with the tendency to think about the need and the demand for energy and for the services it makes possible as something that exists as a social and economic 'fact' and that is, at any one point in time, non-negotiable. Various authors argue that people have certain 'basic' needs, and that these underpin an also basic level of energy demand (Day, Walker and Simcock, 2016; Gough, 2017). Much of this literature supposes that basic needs (for nutrition, shelter, clean water, education, thermal comfort, a non-threatening environment, etc.) are universal. Exactly how these needs are met changes over time, but the contention is that there are certain unwavering requirements and that these account for some, but perhaps not all, energy demands (de Decker, 2018).

In practice, interpretations of exactly what these needs are continue to evolve. For example, services that are now thought to be essential in some societies are considered luxuries, or not considered at all in other countries and contexts. While this is obvious, discussions of entitlement are usually bound in time and space, meaning that they rarely engage with underlying processes of change and variation. Instead, interpretations of acceptable standards of living are treated as matters of normative judgement. From this perspective the challenge is to ensure that people have access to 'necessary' services and resources, whatever these might be. This leads to the conclusion that the amount of energy needed in any one society is that required to enable everyone to participate effectively in it, whatever that entails. Again, these approaches situate demand as something that exists and that is largely independent of technologies and systems of supply thorough which needs are constituted and met.

Ironically, the language of 'demand side management' also takes demand (in the sense of the underlying services and practices that depend on energy consumption) for granted. This concept refers to measures and steps taken to reduce the amount of energy people use. Some forms of demand management are intended to limit consumption at particular times of day. Others, including improvements in energy efficiency, are about minimising the resources needed to deliver specific services. These strategies suppose that aspects of energy consumption can be manipulated and influenced by designers, suppliers and policy makers. For example, real time pricing is intended to encourage consumers to turn off 'unnecessary' appliances, or switch the heating or air conditioning down for a few hours to reduce peak load.

In this context, reference to the 'demand side' simply means that measures are focused on people who use energy rather than on those who generate and deliver it. Programmes of demand side management consequently target consumers (individuals or organisations), who are thought capable of increasing, or decreasing energy consumption (within limits) at will. There is considerable support for such approaches. For example, the European Commission states that 'European households and businesses should be enabled to lower their bills and obtain further benefits if they help relieving pressures in the energy system by simply adapting their demand' (European Commission, 2013, p. 2). As this focus on individual action suggests, 'demand side' measures are not intended to transform the range and character of energy services on which demand depends or to have a major or lasting impact on what people do: they are essentially about trimming energy consumption while maintaining 'normal' levels and standards of service.

The result is an odd situation in which reducing energy consumption is an essential element of climate change policy, and at the same time, the extent and character of 'demand' (what is energy for?, how does this change?) is rarely a topic of analysis and debate. For example, the UK's recent clean growth strategy (HM Government, 2017) does not explicitly discuss demand at all. Instead, the focus is on methods of meeting present 'needs' more efficiently and with fewer carbon emissions than might otherwise be the case. In other work, including that by the UK's Committee on Climate Change, increases in population, growth in GDP, and anticipated improvements in technological efficiency are factored into estimates of future energy consumption. However, this is essentially a matter of quantifying the future energy costs of enabling present ways of life, the basic parameters of which are expected to remain the same. This is not the only possible approach.

Making demand

The positions considered in this second section do not suppose that energy demand 'exists' as a phenomenon in its own right, aside from policy or from systems and technologies of supply. Instead they start from the common proposition that energy demand is an outcome of the social, infrastructural and institutional ordering of what people *do* (Shove and Walker, 2014). Defined in this way, energy demand has a history (in fact, multiple histories) and is constantly changing in line with the practices on which it depends.

For example, the energy 'needed' for transport and mobility is closely related to the development of what Mattioli describes as 'car dependent practices' (Mattioli, Anable and Vrotsou, 2016). Working, shopping and visiting friends and family do not always depend on the use of a car, but in some situations material arrangements such as city planning, road infrastructures and systems of public transport mean that cars and driving are indeed required (Shove et al., 2015). The resulting 'demand' for fuel is not to be taken for granted, nor is it an expression of consumer choice. Rather, 'energy demand is inscribed and reproduced through the combinations of practices and infrastructures of which contemporary forms of car dependence are made' (Shove et al., 2015, p. 275).

Although they underpin what people think of as 'needs', arrangements like these are inherently unstable. From a longer-term perspective, 'needs' are always dynamic and always in flux. Thermal comfort is a good example. Significant amounts of energy are used for heating and cooling buildings and for maintaining the increasingly standardised indoor conditions in which people live, work and play. But how much energy is needed, and what is comfort? Meanings and expectations of comfort have evolved alongside technologies of heating and air-conditioning, and along with standardised methods of building design and mechanical engineering. The result is a very specific understanding of 'need' (buildings are now generally designed to deliver around 18–22 °C) arising from correspondingly specific developments in methods of calculation, in judgements about 'normal' forms of clothing and activity, and in building codes and standards (Cass, 2017; Chappells and Shove, 2005; Shove, 2003).

As this example illustrates, meanings of service (here comfort) and related demands for energy are provisionally held in place by a raft of social and technical arrangements, all of which have complicated and contested histories, all of which are open to negotiation and change and none of which are inevitable or natural. This is not an unusual story. The 'need' for the energy used in lighting, cooking or watching TV is established in much the same way.

These different 'end uses' combine to form what we might think of as the total demand for energy. However, this is only part of the picture. The constitution of demand also depends on related forms of generation, distribution and supply, all of which are implicated in the emergence and disappearance of energy-related services and practices. Policy strategies like those of 'predict and provide' lead to investment in forms of provision (roads, gas networks, electricity grids) that make specific energy-demanding practices possible. More than that, and as Hughes and others have shown, forms of infrastructural development often depend on actively 'making' and not simply meeting pre-existing or 'unmet' demand (Hughes, 1983).

These are not localised or isolated processes. For instance, infrastructures (reliable electricity networks; frozen food distribution systems) and domestic appliances (refrigerators and fridge freezers) are jointly implicated in the development of increasingly global systems of food production and retailing, linked to also changing diets and methods of cooking, buying and storing food at home (Rinkinen, Shove and Smits, 2017). The result is a globally distributed process of 'demand-making' that operates across many places and practices at once.

Given that these examples illustrate the making of demand and given that demand reduction is an important policy goal, further questions arise about how governments and institutions might act to foster less energy-demanding services and less resource intensive practices. In other words, what, if anything, can be done to change the very basis of demand?

Strategies for demand reduction

As indicated above, energy and climate change policies tend to take 'demand' for granted, supposing either that demand exists ready-made (that it is a non-negotiable 'need'); that it is made by uncontrollable market processes, or influenced by policies in other fields. From this perspective, it is perhaps unrealistic, and perhaps not legitimate to even think about intervening with the aim of changing the nature of demand itself. On the other hand, all policy positions influence the ongoing trajectories of consumption and provision. From this point of view, taking demand for granted is itself an active stance, favouring some forms of investment and making some future social practices more likely than others.

It follows that there is, at least in theory, scope for governments to change tack and to think about opportunities for reducing demand 'at source'. Some methods, like planning cities in ways that limit the need to travel, are already established. Other strategies, like those required to challenge contemporary understandings of comfort are less well-developed (one exception being the Japanese government's Cool Biz programme that changed dress codes and reduced the need for air conditioning). Meanwhile, technology-driven goals of increasing reliance on renewable sources of power, or of substituting electric for petrol-powered cars, are set to shift features of infrastructure and provision. Whether intended or not, interventions like these are likely to have some impact on the detail of daily life and hence on the extent and the timing, and not just the (decarbonised) form of energy demand that follows.

Equally important, policy makers involved in making and implementing strategies in areas including health care, economic policy and more, are also involved in changing infrastructures and related systems of provision and practices – all of which matter for energy demand. Recognising that energy demands are defined and shaped by a wide range of non-energy policies leads to the obvious, and also simple, suggestion that non-energy policies might be deliberately mobilised as instruments of demand reduction.

In conclusion, if we define demand as the amount of energy consumed in delivering an already specified service (or set of services) demand reduction is likely to depend on forms of efficiency, on methods of demand side management, and on means of decarbonising supply. All these strategies focus on meeting 'needs' as effectively as possible and on doing so in discrete sectors like those of transport, buildings, industry and so forth.

On the other hand, if we define demand as an outcome of the many services and practices that energy systems and technologies make possible, demand reduction is about steering the constitution and transformation of what people do. From this point of view, demand reduction depends on reconfiguring services and social practices, not on intervening in different sectors, one at a time, and not on progress in efficiency either. For the moment, it is not clear when, whether or how this more extensive interpretation of demand might take hold, or which government departments or policy actors might take the lead.

In the meantime, and in the short term, this chapter provides an important reminder of the fact that there are very different ways of conceptualising 'demand', and that these have practical consequences for how agendas and priorities are established and pursued in research and policy.

Further reading

Shove, E. and Walker, G. (2014). What is energy for? Social practice and energy demand, *Theory, Culture & Society*, 31(5), pp. 41–58.

References

Asif, M. and Muneer, T. (2007). Energy supply, its demand and security issues for developed and emerging economies. *Renewable and Sustainable Energy Reviews*, 11(7), pp. 1388–1413.

Cass, N. (2017). Energy-related standards and UK speculative office development. *Building Research & Information*, 46(6), pp. 615–635.

Chappells, H. and Shove, E. (2005). Debating the future of comfort: Environmental sustainability, energy consumption and the indoor environment. *Building Research & Information*, 33(1), pp. 32–40.

Day, R., Walker, G. and Simcock, N. (2016). Conceptualising energy use and energy poverty using a capabilities framework. *Energy Policy*, 93, pp. 255–264.

EUED (2016). EUED Centres strategy. Retrieved from www.eueduk.com/wp-content/uploads/2016/12/EUEDCentresStrategyMay16-1.pdf [accessed 1 March 2018].

European Commission (2013). Incorporating demand side flexibility, in particular demand response, in electricity markets. Retrieved from https://ec.europa.eu/energy/sites/ener/files/documents/com_2013_public_intervention_swd07_en.pdf [accessed 1 March 2018].

Gough, I. (2017). *Heat, greed and human need: Climate change, capitalism and sustainable wellbeing*. Cheltenham: Edward Elgar Publishing.

HM Government (2017). The Clean Growth Strategy. Retrieved from www.gov.uk/government/uploads/system/uploads/attachment_data/file/651916/BEIS_The_Clean_Growth_online_12.10.17.pdf [accessed 1 March 2018].

Hughes, T. P. (1983). *Networks of power: Electrification in Western society, 1880–1930*. Baltimore, MD: Johns Hopkins University Press.

Mattioli, G., Anable, J. and Vrotsou, K. (2016). Car dependent practices: Findings from a sequence pattern mining study of UK time use data. *Transportation Research Part A: Policy and Practice*, 89, pp. 56–72.

Oxford Dictionaries (2018). Demand. Retrieved from https://en.oxforddictionaries.com/definition/demand [accessed 15 March 2018].

Rinkinen, J., Shove, E. and Smits, M. (2017). Cold chains in Hanoi and Bangkok: Changing systems of provision and practice. *Journal of Consumer Culture*. Online first. doi: 10.1177/1469540517717783.

Shove, E. (2003). *Comfort, cleanliness and convenience: The social organization of normality*. Oxford: Berg.

Shove, E. and Walker, G. (2014). What is energy for? Social practice and energy demand. *Theory, Culture & Society*, 31(5), pp. 41–58.

Shove, E., Watson, M. and Spurling, N. (2015). Conceptualizing connections: Energy demand, infrastructures and social practices. *European Journal of Social Theory*, 18(3), pp. 274–287.

3

ENERGY SERVICES

Janine Morley

Do squirrels distinguish between sources of energy (the nut) and the services that energy makes possible (sustenance)?

Introduction

It has long been recognised that energy is not consumed for its own sake but for the services that it provides. Although the majority of energy research and policy remains firmly focused on energy itself: as a resource that is measured in units like kWh, litres or cubic meters and that is knowingly and rationally acquired and used, there is growing interest in 'what energy is for'. Within this literature, concepts of service are often taken for granted or used inconsistently (Fell, 2017). This chapter shows that a focus on energy services can help develop and strengthen understandings of energy demand that go beyond those rooted in a view of energy as a resource.

Energy as a resource: A common but limited perspective

It seems obvious that people use energy (Janda, 2011) and that in order to manage or reduce energy demand, they need to cut back on their consumption. This can be achieved in a number of ways: by installing energy efficient boilers, heat pumps, insulation and renewable generation in homes and workplaces (Bartiaux et al., 2011; Fawcett, 2013); by driving more efficiently, buying fuel-efficient vehicles or choosing to cycle instead (Barkenbus, 2010; Gallagher and Muehlegger, 2011). Advocates of measures like these assume that energy is a quantifiable and knowable resource that people manage by exercising choice in how it is used. It follows that enhancing choice and awareness should facilitate reductions in energy consumption because – it is also assumed – people will be motivated to save money, even if not by concerns about the environmental consequences of what they do (Sweeney et al., 2013).

This understanding of energy as a resource underpins the vast majority of policies aimed at managing or reducing energy demand. In the UK, there is an emphasis on improving the efficiency of appliances, new technologies and buildings through technical standards and on persuading consumers to opt for more efficient solutions (Warren, 2014). Across Europe (European Commission, 2009) and internationally the roll-out of smart meters represents another key area of policy. Smart meters are designed to provide 'near real time information on energy use' so that 'you will be able to better manage your energy use, save money and reduce emissions' (HM Government, 2017b).

Such measures suppose 'that individuals act as 'micro-resource managers' weighing up the costs and benefits of consuming resources in accordance with their desires, opinions, values, attitudes and beliefs' (Strengers, 2011, p. 36). Although it is common and may be appropriate some of the time (e.g. when analysing energy supply systems), this view is limited in that it overlooks more basic questions about 'why people use resources, how these 'needs' and 'wants' are constituted and how they are changing within the broader context of everyday life' (ibid.). Shove makes a similar point, noting that 'when energy is in the spotlight, the services it provides are in the shadows' (Shove, 1997, p. 271). If we are to understand how energy demand changes and how significant reductions in demand might be made, we need to move away from a fixation on 'energy' per se, to consider the services that energy provides.

What are energy services?

People do not consume energy in the same way that they eat food or drink water, nor do they use it in the way they use a kettle, a car or a TV set: that is, by directly handling, controlling or interacting with it. Instead, energy works in the background: to power appliances and vehicles or to provide lighting, heating, ventilation and so on. This means that people do not demand or benefit from energy as such, but from the services that it makes possible. Accordingly, energy services are usually defined as the useful work that energy does and the benefits this provides.

This distinction between energy and service dates back to at least the 1970s when, in arguing for a more strategic approach to planning electricity infrastructure, Amory Lovins (1976) noted that the qualities of supplied energy did not matter to consumers so long as their 'needs' were still met:

> People do not want electricity or oil, nor such economic abstractions as 'residential services,' but rather comfortable rooms, light, vehicular motion, food, tables, and other real things.
>
> *(Lovins, 1976, p. 78)*

The point is that 'real things' can be achieved through other means than electricity, which was increasingly being used as a default source of power.

The term 'energy service' is then taken up by Reister and Devine as something that is 'measured in units of work, of heat at various temperatures, etc., but these quantities are merely surrogates for measures of the satisfaction experienced when human wants are fulfilled via the direct use of energy' (Reister and Devine, 1981, p. 305). Distinguishing the use of energy from the 'measure(s) of service' it provides, allowed these authors to show how technological changes in the means of provision affect the overall cost of energy services. Since then, other researchers have found it helpful to distinguish between the costs of energy as a resource and the costs of the services it provides. This is, for instance, important when thinking about the consequences of improving efficiency. By reducing the energy required to deliver a given service, such measures also reduce the cost of that service, and may increase the demand for it (Berkhout, Muskens and Velthuijsen, 2000; Herring and Roy, 2007) (see Chapter 5 on rebound).

This begs further questions about what determines the demand for services like heating or lighting. Is it just a matter of their cost, perhaps relative to incomes? Anthropologists and sociologists suggest that the quantity of heat that people 'buy' is also, and perhaps more significantly, related to social, cultural and collective histories and traditions linked to 'socio-technical change and the co-evolution of infrastructures, devices, routines and habits' along with the 'inter-dependent practices of producers, providers, utilities and governments' (Wilhite et al., 2000, p. 118). Although it is important to think about heating and lighting, and not just energy, there is a further step to take. As detailed below, lighting and heating are not the only kinds of service that energy makes possible (Shove, 2003).

In reviewing the ways in which energy services have been defined, Fell (2017) notices that the term is used in two different senses. The first refers to situations in which electricity or fuels are converted into more 'useful' formats, such as lighting, heating, motion, sound and combinations of these in the functioning of appliances like washing machines and computers. Fell refers to these outputs as 'energy services' (ibid.).

The second common meaning refers to the benefits that 'energy services' provide or facilitate, and thereby their usefulness. This refers back to the 'real things' that are demanded directly or indirectly in the course of everyday life: things like comfort, getting to work, sending an email or going on holiday, which Fell (2017) describes as 'end services or states'. Thus, heating (energy service) is undertaken as a means to achieve thermal comfort (end service), and lighting (energy service) is used for the purpose of seeing at night (end service).

It is vital to recognise the difference between these two kinds of service. For example, while it may help, ensuring that room temperature in a building falls within a certain (limited) range does not deliver or guarantee comfort. The room may be unoccupied; or those who are in it might find it too warm, depending on how they are dressed. That is to say, 'energy services' like heating are not inherently useful. In Fell's (2017) terminology, it is 'end services or states' that are actually closer in meaning to the general understanding of energy services as the *useful* work and/or benefits that energy provides.

Despite the centrality of 'end services', they remain poorly conceptualised and often overlooked. In particular, questions of comfort or of what journeys are 'for', tend to be neglected by economic analyses of 'service demand'. As Haas et al. (2008) explains in the case of travel, while 'the *actual energy service* is to reach the shop where I can buy a certain product or to reach my office. . . a common and more technical definition of transport energy services are distances travelled' (ibid., p. 4012, emphasis added). Due to the methods used to model and analyse services and demand for them, the purpose of journeys or the characteristics of comfort are often out of sight. While these issues are attended to in other literatures (Mattioli, Anable and Vrotsou, 2016; Nicol and Humphreys, 2009), this is often not the case in the energy field. It is worth thinking further about how such broad 'end services' as 'comfort' or 'mobility' are conceptualised and how they might be included in research and policy making alike.

Energy services, meta-services and energy demand

The concept of final or end services is familiar in the analysis of energy and material flows. Final services are categories of consumption that can be achieved in more and less eco-efficient ways, commonly including communication, illumination, hygiene, sustenance or nourishment, mobility or transport, shelter or structure, and thermal comfort (Cullen and Allwood, 2010; Roelich et al., 2015; Heiskanen and Pantzar, 1997; Baccini and Brunner, 1991). Such services are characterised in various ways (Heiskanen and Pantzar, 1997) but there is a tendency to interpret them as enduring and universal types of 'need', desire or function that are always present in some form and that must be satisfied in some way. Such an 'excessively functional view' (ibid., p. 424) fails to address the 'socio-historical dimensions' of categories that are themselves changing and culturally varied.

Not surprisingly, anthropologists and sociologists often work with a broader, social and historical notion of services. For example, Wilhite et al. (1996) investigated cross-cultural variations in the meaning and achievement of what they called 'cultural services' such as cosiness and cleanliness, which just happen to depend on energy. Shove (2003) also developed an account of sociotechnical change in everyday forms of consumption in which similar services, like comfort and cleanliness, were central to the analysis. Defined as 'composite accomplishments generating and sustaining certain conditions and experiences' (ibid., p. 165), there is a resemblance between Shove's use of the term 'service'; and that of 'end services or states' (Fell, 2017).

However, to distinguish them from functional understandings I introduce the term 'meta-service' to describe these more expansive, composite 'services' to which energy services contribute. What does the concept of 'meta-services' add to understandings of energy demand? The following three features are especially important (Morley, 2018).

More-than-energy services

Firstly, meta-services like comfort, cleanliness, entertainment or mobility are useful or beneficial conditions, experiences and achievements that depend upon *more* than energy alone. Sometimes energy is not required; often human labour or inputs of some kind are. Meta-services almost always depend on other products, appliances and infrastructures, many of which may not consume energy directly themselves (such as roads, building structures and furniture). The concept of meta-services thereby calls attention to how 'suites of technologies operate together' (Shove, 2003, p. 60) in a 'blend of method, meaning and hardware' (ibid., p. 166) to 'co-constitute[s] the collective conventions of everyday life' (ibid., p. 60). One consequence is that it is not just energy-consumers who are responsible for the shaping of meta-services and the demand for energy services they entail. Many different parties are involved in provisioning, structuring and defining meta-services.

For instance, the fashion industry which supplies clothing and shapes our understanding of what is appropriate to wear is implicated in the achievement of thermal comfort (Morley, 2014).

No fixed or universal 'needs'

Secondly, it is tempting to interpret services such as comfort or mobility as fixed and universal categories of 'need' to which energy services (like heat or transport) contribute in more or less efficient ways. But a key quality of meta-services is that they vary from culture to culture and change over time. This means that what we understand, recognise and physically experience as thermal comfort today, may have been very different in the past; and might not even have existed as a distinctive concept in its own right. In other words, comfort is not simply a pre-existing end for which heating and cooling technologies were developed and thereby became immediately useful. Rather, 'something must have changed in everyday life for these new technologies to *become of use* to large numbers of people' (Kuijer and Watson, 2017, p. 78, emphasis in original).

Meta-services change when their components change

The third key point is that meta-services change when the means of achieving them change (Heiskanen and Pantzar, 1997). For instance, concepts of comfort are conditioned by the technologies that sustain and enable them. It is not uncommon today for people to attribute their own experience of indoor comfort to heating or cooling systems, 'forgetting' that what they are wearing, and their levels of activity and health also contribute. The tendency to equate comfort with the operation of heating and cooling technologies is something that has evolved over time, and that has evolved with those same technologies. Similarly, new 'needs' for communication emerge and co-evolve alongside new 'means', as in the way that internet technologies have incrementally become 'necessary' for most households (Walker, Simcock and Day, 2016).

One consequence is that apparently simple substitutions between different 'means', including more energy-efficient technologies, may lead to shifts in how meta-services are understood and experienced. For instance, air-to-air heat pumps are more energy-efficient but they tend to be used in different ways compared to central heating or direct heating, including for air conditioning in the summer (Gram-Hanssen, Christensen and Petersen, 2012).

In sum, the demand for energy services depends on how meta-services are organised. By focusing on meta-services as more encompassing formations of convention, experience and means of provision, in which energy services like lighting and heating are usually only one among many components, we can better recognise the dynamic and collectively distributed nature of demand. As described above, changes in heating-related energy use are not only a question of how people perceive and manage heat, but also of the wider arrangements in which such experiences transpire: what type of clothing is worn and how indoor spaces are designed and used.

What does this analysis of meta-services imply for demand management policy and intervention?

Managing meta-services to reduce energy demand

When energy is viewed as a resource, energy demand is simply interpreted as the volume of energy used (in other words it is equivalent to energy use itself). Demand management and reduction strategies therefore focus on reducing such quantities, and the most common approach is through energy efficiency (see Chapter 4). But when energy as a commodity is put to one side, and the services that energy provides are in the spotlight, demand management is a matter of *managing or reducing the demand for services*.

There is some interest in modelling the impact of changes in service demand on carbon reduction policies (Haas et al., 2008; Kesicki and Anandarajah, 2011; Kainuma et al., 2013; Fujimori et al., 2014). However, exercises like these tend to focus on economic factors such as the price of services or incomes. Beyond this, the possibility for interventions in service demand, as part of energy or wider carbon reduction policies, does not receive much attention.

There could be several reasons for this, including the way that government departments and energy companies are organised, and what kinds of intervention are considered to be within their remit. But that remit is not fixed, and there may be a case for extending and reformulating it in order to achieve the long term and sizeable reductions in carbon emissions required by climate change targets. The analysis set out here suggests that there is no need to worry that service-demand reduction equates to reductions in levels of wellbeing or quality of life; which can be one concern. This is because meta-services are not the same as energy services. Thus, changes in quantifiable levels of energy services (e.g. room temperature), which is the goal of service-demand reduction, do not translate directly into equivalent changes in meta-services (e.g. comfort). In fact, some reductions in energy services might bring about improvements in meta-services: such as when over-heating in buildings occurs or long-distance commuting declines. The point is that levels of heating, lighting or travel are not themselves synonymous with levels of well-being. There may be some important dependencies, but also considerable scope for re-organisation.

What would it mean to take meta-services as a point of intervention for policy makers charged with the task of reducing energy demand? In theory forms of 'meta-service management' would focus on how meta-services are organised and thus the demand for energy services that they generate. This is not so far away from current discussions about how services are provided and about the potential shift to service-based business models. One complication is that energy policy makers are not the only actors involved.

If as suggested above, meta-services consist of composite arrangements of products, systems, understandings and practices, re-shaping them in less energy-intensive ways will likely involve a number of different parties. This means extending efforts beyond end-consumers, since they are not the only ones involved in shaping meta-services like comfort or entertainment, and also beyond energy actors and explicitly energy-related forms of governance. Developing meta-services in ways that reduce energy demand means identifying the parties who are *already* shaping relevant experiences, conventions and the means of providing for them, and devising ways of reconfiguring their involvement.

For example, the producers of energy-consuming appliances are already required to (re)design their products to meet efficiency standards. But these producers also promote the use of such products in the first place, and when they do so they are potentially changing and reinforcing particular concepts of meta-service, such as what the home or family life should be like. Such shifts could well be in more energy-intensive directions.

Bringing such parties into the project of reformulating meta-services (to radically reduce carbon emissions) is likely to be challenging. Broadly it would mean tracing and emphasising the indirect implications of commercial enterprise and policy decisions for overall energy demand, perhaps as a point of public, social, and ethical concern. One approach to this might be to extend the kind of indirect environmental impacts (or externalities) that are addressed in corporate social responsibility evaluations. Another might be through developing industry-wide initiatives, potentially on a voluntary basis and organised by third sector organisations. For instance, rather like the involvement of supermarkets in the issue of food waste (Evans, Welch and Swaffield, 2017), large fashion retailers and clothing manufacturers could become involved in discussions about sustainable or seasonal clothing and the energy demand that clothing styles imply for buildings. Governments may also have a role in funding or leading initiatives like the Cool Biz programme in Japan, which changed conventions of office wear thereby reducing the 'need' for air conditioning and for the energy demand associated with it (Shove, 2014).

Another route is to focus more explicitly on the delivery and definition of 'services'. Such strategies already feature in some areas of energy-related policy. Service-based business models which focus on selling services, rather than products or commodities, are often invested with the promise of reducing energy demand. This is because the financial incentives to improve energy efficiency are shifted to providers when customers buy services (for instance warm buildings or cloud-based computing services), rather than resources. Policies in the UK support the development of energy service companies (ESCos) and their extension beyond their current, limited role in large organisations (HM Government, 2017a). While this seems like a move in the right direction, there are some risks involved. For instance, if utility companies attempt to *sell* comfort, cleanliness, entertainment or communication, this could entrench existing and energy-intensive interpretations of these meta-services, rather than challenging them and reducing demand for the energy services on which they depend. For instance, comfort as provided by a utility company is likely to be defined by the energy service they can actually deliver (heating or cooling), occluding the wide range of other means of achieving comfort – such as wearing different clothing. In other words, the shift to service-based business models does not inherently challenge 'meta-services' or reduce levels of energy service demand.

Conclusion

To conclude, policies and strategies that focus on energy as a resource are inherently limited. To be fixated on energy, and even on energy services, excludes serious consideration of the social and collective arrangements (meta-services) to which these forms of provision contribute. Not only does this run the risk of misunderstanding how and why energy demand is changing, it also obscures opportunities for managing and reducing energy demand. Exploiting this potential depends, at a minimum, on recognising the roles that diverse policies, companies and practices play in shaping meta-services, now and in the future.

Further reading

Fell, M. J. (2017). Energy services: A conceptual review. *Energy Research & Social Science*, 27, pp. 129–140.

Morley, J. (2018). Rethinking energy services: The concept of 'meta-service' and implications for demand reduction and servicizing policy. *Energy Policy*, 122, pp. 563–569.

Shove, E. (2003). *Comfort, cleanliness and convenience: The social organization of normality.* Oxford: Berg.

Wilhite, H., Shove, E., Lutzenhiser, L. et al. (2000). The Legacy of Twenty Years of Energy demand management: We know more about individual behaviour but next to nothing about demand. In E. Jochem, J. Sathaye and D. Bouille (eds), *Society, behaviour, and climate change mitigation.* Dordrecht: Kluwer Academic Publishers, pp. 109–126.

References

Baccini, P. and Brunner, P. H. (1991). *Metabolism of the Anthroposphere.* Berlin: Springer-Verlag.

Barkenbus, J. N. (2010). Eco-driving: An overlooked climate change initiative. *Energy Policy,* 38, pp. 762–769.

Bartiaux, F., Gram-Hanssen, K., Fonseca, P., et al. (2011). A practice-theory based analysis of energy renovations in four European countries. *Proceedings of 2011 ECEEE Summer Study,* pp. 67–78. Retrieved from http://vbn.aau.dk/files/56120898/1_204_Bartiaux.eceee2011.pdf [accessed 22 November 2018].

Berkhout, P. H. G., Muskens, J. C. and Velthuijsen, J. W. (2000). Defining the rebound effect. *Energy Policy,* 28, pp. 425–432.

Cullen, J. M. and Allwood, J. M. (2010). The efficient use of energy: Tracing the global flow of energy from fuel to service. *Energy Policy,* 38, pp. 75–81.

European Commission (2009). Directive 2009/72/EC of the European Parliament and of the Council of 13 July 2009 concerning common rules for the internal market in electricity and repealing Directive 2003/54/EC. Retrieved from https://eur-lex.europa.eu/legal-content/EN/ALL/?uri=celex%3A32009L0072 [accessed 22 November 2018].

Evans, D., Welch, D. and Swaffield, J. (2017). Constructing and mobilizing 'the consumer': Responsibility, consumption and the politics of sustainability. *Environment and Planning A,* 49, pp. 1396–1412.

Fawcett, T. (2013). Exploring the time dimension of low carbon retrofit: Owner-occupied housing. *Building Research & Information,* 42(4), pp. 477–488.

Fell, M. J. (2017). Energy services: A conceptual review. *Energy Research & Social Science,* 27, pp. 129–140.

Fujimori, S., Kainuma, M., Masui, T., et al. (2014). The effectiveness of energy service demand reduction: A scenario analysis of global climate change mitigation. *Energy Policy,* 75, pp. 379–391.

Gallagher, K. S. and Muehlegger, E. (2011). Giving green to get green? Incentives and consumer adoption of hybrid vehicle technology. *Journal of Environmental Economics and Management,* 61, pp. 1–15.

Gram-Hanssen, K., Christensen, T. H. and Petersen, P. E. (2012). Air-to-air heat pumps in real-life use: Are potential savings achieved or are they transformed into increased comfort? *Energy and Buildings,* 53, pp. 64–73.

Haas, R., Nakicenovic, N., Ajanovic, A., et al. (2008). Towards sustainability of energy systems: A primer on how to apply the concept of energy services to identify necessary trends and policies. *Energy Policy,* 36, pp. 4012–4021.

Heiskanen, E. and Pantzar, M. (1997). Toward sustainable consumption: Two new perspectives. *Journal of Consumer Policy*, 20, pp. 409–442.

Herring, H. and Roy, R. (2007). Technological innovation, energy efficient design and the rebound effect. *Technovation*, 27, pp. 194–203.

HM Government (2017a). Registered energy service providers. Retrieved from www.gov.uk/government/publications/registered-energy-service-providers. [accessed 22 November 2018].

HM Government (2017b). Smart meters: A guide. Retrieved from www.gov.uk/guidance/smart-meters-how-they-work [accessed 22 November 2018].

Janda, K. B. (2011). Buildings don't use energy: People do. *Architectural Science Review*, 54, pp. 15–22.

Kainuma, M., Miwa, K., Ehara, T., et al. (2013). A low-carbon society: Global visions, pathways, and challenges. *Climate Policy*, 13, pp. 5–21.

Kesicki, F. and Anandarajah, G. (2011). The role of energy-service demand reduction in global climate change mitigation: Combining energy modelling and decomposition analysis. *Energy Policy*, 39, pp. 7224–7233.

Kuijer, L. and Watson, M. (2017). 'That's when we started using the living room': Lessons from a local history of domestic heating in the United Kingdom. *Energy Research & Social Science*, 28, pp. 77–85.

Lovins, A. (1976). Energy strategy: The road not taken. *Foreign Affairs*, 1, pp. 65–96.

Mattioli, G., Anable, J. and Vrotsou, K. (2016). Car dependent practices: Findings from a sequence pattern mining study of UK time use data. *Transportation Research Part A: Policy and Practice*, 89, pp. 56–72.

Morley, J. (2014). Diversity, dynamics and domestic energy demand: A study of variation in cooking, comfort and computing. Doctoral thesis, Lancaster University. Retrieved from www.research.lancs.ac.uk/portal/en/publications/-(121c3acced34-412d-b90e-5e61b8dea856).html [accessed 26 February 2019].

Morley, J. (2018). Rethinking energy services: The concept of 'meta-service' and implications for demand reduction and servicizing policy. *Energy Policy*, 122, pp. 563–569.

Nicol, J. F. and Humphreys, M. A. (2009). New standards for comfort and energy use in buildings. *Building Research & Information*, 37, pp. 68–73.

Reister, D. B. and Devine, W. D. (1981). Total costs of energy services. *Energy*, 6, pp. 305–315.

Roelich, K., Knoeri, C., Steinberger, J. K., et al. (2015). Towards resource-efficient and service-oriented integrated infrastructure operation. *Technological Forecasting and Social Change*, 92, pp. 40–52.

Shove, E. (1997). Revealing the invisible: Sociology, energy and the environment. In M. Redclift and G. Woodgate (eds), *International handbook of environmental sociology*. Cheltenham: Edgar Elgar, pp. 261–273.

Shove, E. (2003). *Comfort, cleanliness and convenience: The social organization of normality*. Oxford: Berg.

Shove, E. (2014). Putting practice into policy: Reconfiguring questions of consumption and climate change. *Contemporary Social Science*, 9, pp. 415–429.

Strengers, Y. (2011). Beyond demand management: Co-managing energy and water practices with Australian households. *Policy Studies*, 32, pp. 35–58.

Sweeney, J. C., Kresling, J., Webb, D., et al. (2013). Energy saving behaviours: Development of a practice-based model. *Energy Policy*, 61, pp. 371–381.

Walker, G., Simcock, N. and Day, R. (2016). Necessary energy uses and a minimum standard of living in the United Kingdom: Energy justice or escalating expectations? *Energy Research & Social Science*, 18, pp. 129–138.

Warren, P. (2014). A review of demand-side management policy in the UK. *Renewable and Sustainable Energy Reviews*, 29, pp. 941–951.

Wilhite, H., Nakagami, H., Masuda, T., et al. (1996). A cross-cultural analysis of household energy use behaviour in Japan and Norway. *Energy Policy*, 24, pp. 795–803.

Wilhite, H., Shove, E., Lutzenhiser, L., et al. (2000). The legacy of twenty years of energy demand management: We know more about individual behaviour but next to nothing about demand. In E. Jochem, J. Sathaye and D. Bouille (eds), *Society, behaviour, and climate change mitigation*. Dordrecht: Kluwer Academic Publishers, pp. 109–126.

PART II
Characteristics

4

ENERGY EFFICIENCY

Elizabeth Shove

Ostriches are said to put their heads in the sand when faced with danger. Whether ostriches really do this or not, the image of wilfully excluding complicating or extraneous considerations resonates with the main themes of this chapter.

Introduction

The goal of increasing the energy efficiency of products, buildings, and processes has figured in national and international policies from the 1970s onwards, and it remains at the heart of national and international responses to climate change. Energy efficiency is generally defined as a matter of achieving more or the same service but for less energy: a broad ambition that is, at first sight, difficult to fault. The 'savings' attributed to energy efficiency are often significant. For example, the International Energy Agency (IEA) estimates that 'Globally, efficiency gains since 2000 prevented 12% more energy use than would have otherwise been the case in 2017' (IEA, 2018a). Given figures like these, it is not surprising that energy efficiency is such a high priority.

To give just a few examples, the UK's Committee on Climate Change identifies two principal ways of reducing carbon emissions: energy efficiency is one, decarbonising supply is the other (Committee on Climate

Change, undated). The European Commission's 2030 climate and energy framework establishes the target of increasing 'energy efficiency' by 27 per cent compared with the 'business as usual' scenario (European Commission, 2014). Other organisations have similar ambitions. The European Council for an Energy Efficient Economy is committed to keeping 'energy efficiency first' in the line of responses to climate change (European Council for an Energy Efficient Economy, undated; the American Council for an Energy Efficient Economy has a similar ambition) and the IEA asserts that 'Energy efficiency is key to ensuring a safe, reliable, affordable and sustainable energy system for the future', also suggesting that 'It is the one energy resource that every country possesses in abundance and is the quickest and least costly way of addressing energy security, environmental and economic challenges' (IEA, 2018b).

These ideas justify investment in the production of more efficient appliances (from fridge-freezers, through to boilers, outdoor patio heaters, air conditioning systems and light bulbs), and of measures like those that improve building insulation. All other things being equal, initiatives of this kind really do reduce the energy that might otherwise be used, but they also have a range of other, unintended and often undesirable consequences. Running against conventional wisdom, this fable highlights the limitations and the potentially counterproductive side effects of an unthinking pursuit of energy efficiency, whether alone or as part of a broader 'whole systems' approach.

What is wrong with energy efficiency?

There are two main critiques of energy efficiency. One focuses on the problem of 'rebound', the other on the interpretations of normality and service that are embedded in measures of efficiency and in related policies.

The most widely discussed problem with efforts to improve energy efficiency is that the benefits (in terms of reduced carbon emissions, and energy) might not be fully realised if people 'spend' the savings that follow in ways that increase energy demand in the system or society as a whole. This is known as the 'rebound' effect (see Chapter 5). The classic example is of a household that acquires a more efficient car but puts the money saved on fuel towards the cost of a foreign holiday. Similarly, people who install a more efficient boiler often 'take back' some of the benefit by keeping their home warmer than it was before – again reducing the extent to which efficiency 'savings' are actually realised (Rees, 2009, p. 304; Hamilton et al., 2016).

Rebound effects relate to the Jevons paradox, named after William Stanley Jevons, an economist who made the link between increased technological efficiency (in his case regarding the use of coal), reduced costs and increased consumption. There are differences of opinion about the scale of such effects, and ongoing arguments about the macro- and micro- and the longer and shorter-term consequences of efficiency (Herring, 2006, p. 10).

To date, the more fundamental, long-term problem represented by the Jevons paradox – that increased efficiency enables increased consumption –receives little explicit attention within policy agendas that are quite tightly bound in space and time. Instead, efforts to identify and quantify rebound effects focus on the trade-offs and choices made usually by households, but sometimes by cities or by entire nations. Estimating the magnitude of rebound is a complicated task, and one that has received considerable attention over the years. For the most part, this effort has been driven by the goal of *improving* rather than totally overhauling or abandoning efficiency policy (Sorrell, 2015).[1] In other words, discussions about rebound or 'backfire' tend to be about how much allowance should be made for the possibility that efficiency measures might not work (in aggregate) quite as well as expected.

A second and in many ways more powerful charge is that programmes of efficiency unwittingly contribute to increases in energy demand in that they reproduce and foster specific, and often resource intensive understandings of 'service' (see Chapter 3). To understand how this works we need to take a step back. As is obvious, efficiency is not the same as consumption. For example, it is possible to buy a very large fridge freezer that uses twice the energy of a much smaller model, but that is rated as being extremely efficient compared with others in its class. Measures of energy efficiency involve comparisons between appliances (or systems) that are identical in all respects other than the amount of energy they use. This makes sense. Comparing the energy consumption of fridge-freezers that offer different facilities (e.g. that are larger, or have ice dispensers, etc.) would reveal nothing about the relative efficiency of each. Judgements of efficiency consequently depend on first establishing and then holding steady a specification of 'equivalent' performance (Shove, 2017).

Rees and others argue that since efficiency is a matter of delivering *the same or more service* but for less energy, efficiency measures persistently 'reproduce the status quo by other means' (Rees, 2009, p. 304). As a result, and as Thomas et al. conclude, discourses and measures of efficiency consequently fail to halt and might even foster transformations in expected needs and interpretations of service which underpin increases in demand (Thomas et al., 2015). The next section details some of the routes through which seemingly obvious, seemingly technical measures have this effect.

How programmes of energy efficiency reproduce the status quo

Describing and measuring energy efficiency depends on developing agreed methods of (a) knowing and quantifying energy input and of (b) defining and determining what counts as 'the same or more' service. Further decisions are needed about how objects of efficiency are to be (c) specified and (d) compared over time. At each stage, different kinds of assumptions are necessarily involved.

Measuring energy

Over the last few centuries, units like joules, kWh or million tonnes of oil equivalent (Mtoe) have taken the place of previously diverse, contextually situated methods of representing energy in terms of horse power, manpower or candle power (Shove, 2017). This is an important development in that generic metrics make it possible to aggregate and compare energy use and to characterise the efficiencies of very different commodities and entities in the same terms. Such measures have the further effect of constituting energy as an all-purpose resource, rather than as something that is generated and consumed in ways that are highly contingent, variable and historically specific. By abstracting energy from the exceptionally diverse social processes and practices (keeping warm, driving to work, doing the laundry and so forth) in which it is, in fact, embedded, estimates of the amount of energy 'avoided' or not used (because of energy efficiency) tell us nothing about what energy is used for or how uses of energy evolve over time. As Lutzenhiser explains, reliance on standardised units and measures helps constitute a realm of engineering and policy analysis in which problems of energy are separated from processes of social and cultural change (Lutzenhiser, 2014).

Establishing equivalent service

There are many ways of describing the services provided by a home, a room or an appliance but as mentioned above, judgements of relative efficiency depend on a single, unambiguous method of specifying and standardising meanings of service. The usual approach is to select one quality – such as the light output of a bulb, measured in lumens – and to design solutions that deliver the same or more light for less energy (Diamond and Shove, 2015). With service characterised in these terms, other aspects, like weight, cost or the quality of light produced are excluded from the efficiency equation. This is relevant in that technological developments that improve performance on one dimension often have consequences for other features, meaning that more efficient solutions are almost always *different* – in some respects – from those with which they are compared. For example, low-e windows are thermally more efficient but typically reduce the transmission of daylight and of sound, and the fading of fabrics, meaning that rooms also appear darker. In these and in other such situations, establishing equivalence depends on foregrounding certain characteristics over others and fixing these as indicators of service in relation to which relative efficiencies are assessed. This is one of the routes through which assumptions about need and service become taken-for-granted and carried forward. Others relate to the terms in which objects of efficiency are defined and framed.

Bounding objects of efficiency

There are many possible objects, systems or entities that can be evaluated and compared in terms of their relative efficiency. It is, for example, possible to compare the efficiency of one central heating boiler with another, or to rate the efficiency of different components – like the motor of a dishwasher or the engine of a car. Alternatively, one might compare the relative efficiencies of entire houses or complete cars. Either way, judgements about efficiency depend on slicing objects – whether boilers, engines or houses – out of context and treating them as discrete entities in their own right. While this is a necessary step, it is again one that disguises ongoing and often complex relationships between parts and wholes and the impact these have on changing interpretations of service.

This becomes obvious when thinking about the efficiency of the current Citroen C1 compared with a 2CV from the 1950s. The Citroen C1 does about the same number of miles per gallon (mpg) as the 2CV (de Decker, 2008). Although the C1's engine is much more efficient than that of the 2CV, it is used to drive a vehicle that is heavier, that has windows that wind down, and that has all the features one would expect of a car today. The mpg would increase significantly if manufacturers were to put a modern engine inside an old 2CV, but the result would not correspond to what now counts as a 'car'. Because definitions of efficiency depend on bounding objects (a car, an engine, a heating system, a house), they reveal nothing about how the meanings and services expected of such entities develop and interact.

Not surprisingly, efforts to increase efficiency depend on how the 'object' of such attention is defined and bounded. For example, estimates of the efficiency of a house generally focus on the amount of gas and electricity required to maintain a certain indoor temperature, hence the emphasis on insulation and heating systems. Shifting the point of reference and considering the amount of energy used in *keeping people warm*, would bring other 'technologies' into the equation, including clothing. Wearing insulation close to the body is an exceptionally effective method of reducing heat loss, making better use of the body's own energy and thus requiring less additional input for the same warm service. But since it is temperature rather than comfort that counts, 'measures' like those of wearing warmer clothing are routinely excluded from efficiency-related research and policy.

This begs the question of how objects of efficiency come to be constituted as they are. For example, what makes temperature a more suitable 'object' of efficiency than comfort? Part of the answer has to do with the need to simplify and stabilise equivalent units of analysis. Since people dress in very different ways, and since home furnishings are also taken to be matters of personal taste, it makes sense (from an engineering perspective), to take these elements out of the equation, and in the same move build them back in by making standardised assumptions about the thermal properties of curtains, carpets and clothing. Similarly, rather than worrying about the meaning of comfort, it is much easier to establish a 'normal' room temperature and take this as given when calculating the relative efficiencies of different building and insulation materials.

In concentrating on only some problems (the efficient provision of a specified indoor temperature, not the efficient provision of comfort) efficiency programmes marginalise potentially relevant elements (like clothing). In making assumptions about 'normal' operating conditions (including indoor temperatures), such programmes have the double effect of reinforcing contemporary interpretations of what indoor temperatures should be, disregarding the fact that these have changed considerably over time. Analyses of efficiency also tend to focus on certain types of energy at the expense of others. For example, attention is paid to the efficient use of coal, gas, oil or electricity, but not to human labour. The exclusion of human effort is understandable given the importance of establishing 'equivalent service' from the consumers' point of view. But by leaving this out of the equation, the effect is to reproduce tacit understandings of how much physical effort should (or should not) be involved in achieving certain levels of 'service'.

All these judgements have a bearing on the outcome of efficiency calculations, and hence on the steps and interventions that are taken or proposed in response. In this way, programmes of efficiency are 'performative': that is, they are themselves implicated in shaping social processes of interpretation and evaluation.

The time frames of energy efficiency

Some evaluations of efficiency are contemporaneous – as when people compare the relative efficiencies of domestic appliances on sale today. However, there is also interest in tracking improvements in efficiency over time. For example, the IEA's 2015 energy efficiency market report claims that 'Cumulatively, investments since 1990 have generated 256 EJ (6,120 Mtoe) of avoided consumption, with reductions in electricity and natural gas use dominating' (IEA, 2015, p. 17). This does not mean that 1990 is in some absolute sense the dawn of efficiency, but it does mean that previous, also cumulative histories of socio-technical change are out of range. In order to estimate 'avoided energy' the IEA identifies investments over a certain period, and evaluates their impact on bounded objects of efficiency, calculating the benefits with reference to the base case of a chosen year (such as 1990). Picking another point of reference – for instance 1890 – would obviously produce different results. But as this thought experiment suggests, setting 1890 as a baseline is implausible because the services delivered then and now are not at all equivalent.

Discussions of efficiency are simultaneously time bound (they depend on comparison) but also timeless. In purely engineering terms, what matters is the ratio of input to output, never mind when in history or in the future changes in that ratio might occur. In practice, policy analyses tend to focus on the short term, partly because of the need to stabilise definitions of equivalent service and partly to demonstrate effect. Again the point is that selecting one time frame, rather than another, is itself a decision about how to accommodate, or marginalise, social and cultural change in what people do and the energy demands that follow.

To summarise, constituting efficiency as a meaningful topic depends on treating energy as a generic, quantifiable resource and as something that has an ontological reality of its own (Labanca, 2017). Various discursive and methodological moves are needed to disentangle energy and service from the everyday practices in which they are enmeshed. As outlined above, the abstractions involved are in no way neutral: they carry with them both a politics of the present, and a raft of related assumptions about the status of energy and service, and about how both might be specified and measured. They also isolate discourses of efficiency from what are taken to be unrelated market trends in design, manufacturing and consumption. In all these ways, and more, efficiency policies disguise, and in the same move reinforce, their own role in making patterns of energy demand what they are today and in shaping those of the future as well. If we go along with this analysis, it is difficult to avoid the conclusion that energy efficiency policies are implicated in reproducing rather than challenging what are likely to be unsustainable interpretations of 'need' and service (Shove, 2017).

Energy efficiency: Part of the problem or part of the solution?

If the unthinking pursuit of energy efficiency works, via the necessary concept of equivalence of service, to perpetuate and perhaps escalate but never undermine ever more energy intensive ways of life it is unlikely to constitute an effective response to climate change. Instead, the opposite may be true. By holding fast to the view that efficiency linked to present standards of service is self-evidently part of the solution, policy makers may be unwittingly *increasing* the challenges that lie ahead. It is difficult to imagine calculations of relative efficiency that do not depend on selective, simplifying assumptions: some forms of bounding and framing are always going to be required, so do efficiency agendas always or necessarily perpetuate the status quo?

Enabling lower carbon ways of life is evidently not the same as promoting energy efficiency. But if programmes and policies of efficiency are to help limit (rather than foster) more resource intensive practices it is essential to bring these goals much closer together (Harris et al., 2008). Because present approaches do not pay attention to the kinds of services that efficient 'solutions' sustain, they fail to distinguish between forms of efficiency that contribute to escalating interpretations of need and demand, and those that do not. Very large homes, enormous fridge freezers and outdoor patio heaters can all be energy efficient, and so can bicycles, 'tiny homes' and solar cooking stoves. Rather than promoting 'efficiency' across the board, the way forward is to differentiate between 'good' forms which embody interpretations of service that are (apparently) compatible with much lower impact ways of life, and those that are not. Movement in this direction calls for new ways of thinking about how services are constituted, how they change and about what part programmes and technologies – including technologies of efficiency – play in these processes.

Put differently, if energy efficiency is to be part of the solution rather than part of the problem, questions about sufficiency (what is energy used for, how much energy do societies 'need'; how much consumption is compatible with meeting carbon targets) need to come first. Only then is it possible to develop programmes of efficiency that are selectively focused on technologies and services that *do not* meet present needs, and that *do not* deliver equivalent levels of service, but that do facilitate and sustain practices of keeping warm, travelling, cooking, lighting and more that are much less demanding than those we take for granted today. This depends on carefully articulating and questioning the social and cultural assumptions embedded in the techniques and methodologies on which programmes of energy efficiency currently depend, and on challenging the pervasive myth that such programmes are a self-evidently effective component of 'whole systems' energy and climate change policy.

Note

1 Policy proposals designed to temper the effects of rebound typically argue for an energy or carbon tax, suggesting that this would help ensure that the benefits of efficiency are more comprehensively realised, and not diluted or squandered in ways that backfire.

Further reading

Calwell C. (2010). *Is efficient sufficient?* Report for the European Council for an Energy Efficient Economy. Retrieved from www.eceee.org/static/media/uploads/site-2/policy-areas/sufficiency/eceee_Progressive_Efficiency.pdf [accessed 19 November 2018].

Herring H. (2006). Energy efficiency – a critical view. *Energy*, 31, pp. 10–20.

Labanca N. (2017). *Complex systems and social practices in energy transitions: Framing the issue of energy sustainability in the time of renewables.* Cham: Springer International.

Lutzenhiser, L. (2014). Through the energy efficiency looking glass. *Energy Research & Social Science*, 1, pp. 141–151.

Shove, E. (2018). What is wrong with energy efficiency? *Building Research & Information*, 46(7), pp. 779–789.

References

Committee on Climate Change (undated). What can be done. Retrieved from www.theccc.org.uk/tackling-climate-change/reducing-carbon-emissions/what-can-be-done [accessed 19 November 2018]

De Decker, K. (2008). The Citroen 2CV: Cleantech from the 1940s. Retrieved from www.lowtechmagazine.com/2008/06/citroen-2cv.html [accessed 19 November 2018]

Diamond, R. and Shove, E. (2015). Defining efficiency: What is 'equivalent service' and why does it matter? Retrieved from www.demand.ac.uk/05/10/2015/article-defining-efficiency-what-is-equivalent-service-and-why-does-it-matter [accessed 22 Nov 2018].

European Commission (2014). 2030 climate & energy framework. Retrieved from https://ec.europa.eu/clima/policies/strategies/2030_en [accessed 19 November 2018]

European Council for an Energy Efficient Economy (undated). eceee's strategy 2016–2019. Retrieved from www.eceee.org/about-eceee/governance/strategy/strategy-2016-2019 [accessed 19 November 2018]

Hamilton, I. et al. (2016). Energy efficiency uptake and energy savings in English houses: A cohort study. *Energy and Buildings*, 118, pp. 259–76.

Harris, J. Diamond, R., Iyer, M., Payne, C., Blumstein, C. and Siderius, H.-P. (2008). Towards a sustainable energy balance: Progressive efficiency and the return of energy conservation. *Energy Efficiency*, 1, pp. 175–188.

Herring, H. (2006). Energy efficiency – a critical view. *Energy*, 31, pp. 10–20.

IEA (2018a). Energy efficiency 2018: Analysis and outlooks to 2040. Retrieved from www.iea.org/efficiency2018 [accessed 19 November 2018].

IEA (2018b). Energy efficiency: The global exchange for energy efficiency policies, data and analysis. Retrieved from www.iea.org/topics/energyefficiency [accessed 19 November 2018]

Labanca, N. (2017). *Complex systems and social practices in energy transitions: Framing the issue of energy sustainability in the time of renewables.* Cham: Springer International.

Lutzenhiser, L. (2014). Through the energy efficiency looking glass. *Energy Research & Social Science,* 1, pp. 141–151.

Rees, W. (2009). The ecological crisis and self-delusion: Implications for the building sector. *Building Research & Information,* 37, pp. 300–311.

Shove, E. (2017). Energy and social practice: From abstractions to dynamic processes. In N. Labanca (ed.), *Complex systems and social practices in energy transitions: Framing the issue of energy sustainability in the time of renewables.* Cham: Springer International, pp. 207–220.

Sorrell, S. (2015). Reducing energy demand: A review of issues, challenges and approaches, *Renewable and Sustainable Energy Reviews,* 47, pp. 74–82.

Thomas, S., Brischke, L., Thema, J. and Kopatz, M. (2015). Energy sufficiency policy: An evolution of energy efficiency policy or radically new approaches? Paper presented at ECEEE Sumer Study, Toulon. Retrieved from www.eceee.org/library/conference_proceedings/eceee_Summer_Studies/2015/ 1-foundations-of-future-energy-policy/energy-sufficiency-policy-an-evolution-of-energy-efficiency-policy-or-radically-new-approaches [accessed 15 February 2019].

5

REBOUND

Greg Marsden

Frogs are good at jumping – but how does one leap affect the next? The concept of rebound relates to the possibility that the benefits of energy-saving actions might be undermined if people use the resources they save in other undesirable directions.

What is the rebound effect?

Steve Sorrell sets out the following definition of the rebound effect:

> The potential 'energy savings' from improved energy efficiency are commonly estimated using basic physical principles and engineering models. However, the energy savings that are realised in practice generally fall short of these engineering estimates. One explanation is that improvements in energy efficiency encourage greater use of the services (for example heat or mobility) which energy helps to provide. Behavioural responses such as these have come to be known as the energy efficiency 'rebound effect'.
>
> *(Sorrell, 2007, p. v)*

To rebound is defined in the *Oxford English Dictionary* as to 'bounce back through the air after hitting something hard', to 'recover in value, amount, or strength after a decrease or decline' or to 'have an unexpected adverse consequence' on someone.[1] In the energy field, the rebound effect has been developed as an analogy with relations to all three of the above definitions.

First, there is the creation of 'something hard' from which to rebound. In the energy context, this is typically a notion of a hypothetical world in which new technologies are introduced without impacting on the demand for energy services (Llorca and Jamasb, 2017; also Chapters 3 and 4 of this volume). As Llorca and Jamasb state, 'as these improvements [in efficiency] also imply a reduction in the relative cost of the service, they may also lead to some increase in demand for energy services. This increase in energy consumption can, partially or totally, offset the initial expected savings' (Llorca and Jamasb, 2017, p. 99). This anticipated bouncing back constitutes the recovery in value implied in the second part of the definition of rebound. One reason why there can be increased energy consumption relative to engineering expectations is that efficiency improvements make in-use costs cheaper. So, for example, light bulbs that cost less to run may be left on for longer or as cars become cheaper to drive, people may drive further or share fewer lifts. However, there are more aspects of 'rebound' to explore.

A third feature of the rebound analogy has to do with how rebound effects are identified and analysed. The use of the concept of elasticities from the field of micro-economics is a key instrument in the analytical tool box (see Chapter 6). 'Economists typically estimate elasticities of demand (e.g. the marginal change in demand for air conditioning as the operating cost of the air conditioner changes), which can be easily converted into a direct rebound effect' (Gillingham, Rapson and Wagner, 2016, p. 72).[2] Here, the analogy at play relates to rubber or springs (Hooke's law) and the extent to which something changes size when the forces on it change. Demand for goods is said to be inelastic when they are not sensitive to price changes and elastic when they are sensitive. If demand for a service was inelastic then, were that same service to be delivered for less cost, little direct demand change would be seen with greater emphasis on indirect effects.

This chapter takes a critical look at the 'rebound effect' as represented in a number of recent articles on this topic. This is a complex field and there are important and challenging data and definitional issues to be considered to ensure that insights are not used beyond their analytical limits. The rebound literature seeks to overcome some of the limitations of engineering efficiency as a means of informing and evaluating policy. It does so by considering direct and indirect effects of efficiency improvements (usually in specific technologies) and it does so over different time periods (Sorrell and Dimitropoulos, 2008). Such studies are designed to provide useful insights that should prevent policy makers from being drawn into over simplistic and over optimistic assumptions about the effects of technological improvements on the reduction of energy demand. These are a knowable 'unexpected' consequence of efficiency policy as per the third part of the dictionary definition. As well as commenting on these aspects, this chapter foregrounds assumptions that need to be made in order to give rebound meaning.

What is the history of the idea of rebound?

In 1866, the English economist William Stanley Jevons wrote a book entitled *The Coal Question* in which he observed that technological improvements in the efficiency of coal use were not reducing coal consumption but instead increasing it as it became adopted in new industries as well as broadening its application in existing ones. The 'Jevons paradox' is where resource consumption increases when there are technological improvements because of increasing demand. This is the first major conceptual representation of 'rebound' effects even though it is not often labelled as such.

More recently, the so-called Khazzoom–Brookes postulate from the 1980s claims that, if energy prices do not change, cost effective energy efficiency improvements will inevitably increase economy-wide energy consumption above what it would be without those improvements' (Sorrell, 2007); that is, at a broader scale, the improvements 'backfire'. While the theoretical and empirical basis for the Kazzoom–Brookes postulate have been challenged, in so doing, the development of techniques for conceptualising and calculating rebound effects have been substantially developed.

Walnum et al. (2014) find that both micro and macroeconomics have attempted to characterise rebound (Gillingham, Rapson and Wagner, 2016), including distinguishing between direct, indirect, and economic-wide rebound effects. Walnum et al. summarise these as follows:

- A direct rebound effect occurs when improvements in energy efficiency increases the use of products and services. For example, consumers who purchase a new and more fuel efficient car might drive more because it becomes cheaper to drive . . .
- An indirect rebound effect occurs when the money saved on reduced fuel consumption is spent on other energy-intensive goods and services, such as air conditioners or a second car in a household . . .
- The sum of the direct and indirect rebound effects from energy efficiency improvements is termed the economy-wide rebound effect.

(Walnum et al., 2014, p. 9516)

Sorrell (2007) further expands this list to seven types which, beyond direct and indirect include embodied effects (additional energy in materials used in production), service quality effects (such as purchasing larger devices), energy market effects (where sufficient change occurs to affect market prices and consumption of other goods), secondary effects (where change is substantial enough to affect the organisation of multiple markets) and transformation effects (such as the lock in of more car dependent practices).

Taken together, this vast swath of different responses reveals the complexity of how goods and services become embedded in different social and industrial practices over time. For many innovations, several of the effects listed will occur over different timescales. In turn, they may combine with other product changes in other sectors to develop in unforeseen ways. Isolating measurable rebound effects requires the drawing of boundaries and the demarcation of effects which may not really be so isolated. The skill in generating useful insights from any such analysis relates to identifying the most significant effects and being clear about how far interpretation should stretch.

Why is the concept of rebound important?

What matters to the achievement of environmental goals is the change in energy in use and the energy mix from which that is generated. The scientific developments around the notion of rebound derive from a desire to understand the effects of policy on the ground and to explain why technological efficiency measures, particularly on their own, fail to cut energy use either to the extent promised or, in the worst case, why they contribute to net increases in energy use.

Vivanco, Kemp and van der Voet (2016) discuss why, despite a very active academic tradition surrounding rebound, the concept has yet to impact on policy and is rarely part of the policy discourse. Without packages of measures such as tax reforms and product innovation, which together seek to limit the extent to which cost savings bleed in to other areas of consumption, they argue that there will be little progress. Without offering a definitive solution, they suggest that the simplicity of the efficiency argument is alluring to policy makers while the difficulties associated with the consideration of second order effects implied by rebound serves to marginalise their inclusion.

In the review of rebound conducted for the UK Energy Research Centre, Sorrell similarly stated:

> [R]ebound effects are very difficult to quantify, and their size and importance under different circumstances is hotly disputed. Also, rebound effects operate through a variety of different mechanisms and lack of clarity about these has led to persistent confusion.
>
> *(Sorrell, 2007, p. v)*

The tradition of research surrounding rebound has thus derived from a specific purpose: understanding of the impacts of product or service changes on energy in use. In the discussion that follows, the critique focuses not on this purpose, but on the limits to what can be understood given the inevitable simplifications required to calculate what has become known as rebound, and what follows from that.

Which problematic assumptions does the concept of rebound depend on?

One of the core assumptions that underpins rebound is the capacity to describe the potential energy implications of an efficiency innovation of some sort. As discussed in Chapter 4, the very notion of an efficiency improvement requires establishing some sort of equivalence between the product or service that went before and the one that is replacing it. As set out in that chapter, these baselines are not meaningful, or at least only so in particular and narrowly defined terms.

However, engineering assessments of efficiency gains do exist and whatever the merits or problems of using such hypothetical benchmarks, they are used in policy. This is so despite Gillingham, Rapson and Wagner's observation that 'because the microeconomic rebound effect consists of substitution and income effects across *all* goods, an attempt to fully measure the rebound effect would require estimating the substitution and income effects for all goods in the economy – clearly an infeasible task' (Gillingham et al., 2016, p. 73). So, while indirect impacts could be very important they are likely to be diffuse and therefore not considered. Similarly, secondary effects and transformational effects are likely to be analytically intractable.

So, what *is* included in a rebound assessment? Gillingham and colleagues' review suggests that 'most studies ignore the demand for other goods and focus on estimating the price elasticity of demand for the more energy-efficient product' (Gillingham et al., 2016, p. 73). One of the most in-depth examples of rebound calculations found in the literature is by Stapleton, Sorrell and Schwanen (2016), who attempted to identify the rebound effect of improving vehicle engine efficiency. The authors discussed what the right metric would be to reflect rebound, using three different variables of vehicle kilometres per capita, per adult, or per registered licence holder noting that 'estimates of rebound effects may therefore depend on how it is defined' (ibid., p. 315). These metrics require further consideration.

First, demand for travel, measured by distance is only a proxy for the energy service provided by, or as a result of, 'mobility'. The demand for travel is changing slowly over time because of both changes to prices and the efficiency of vehicles, but also because of a vast range of other changes such as land-use and spatial development. If the question relates to a short-run change such as the impacts of a one off 10 pence per litre reduction in fuel duty to the Exchequer in the coming year then these broader trends could perhaps be considered as small relative to the direct effect of the price change. However, where effects will play out over longer periods of time it seems more challenging to assume equivalence of what a vehicle kilometre in say 2017 will allow someone to achieve relative to one in 2047. While techniques can be applied to try to include some of these wider variables (e.g. how urbanisation changes over time) there are so many variables at play that correlate to each other that modelling is difficult (see Stapleton, Sorrell and Schwanen, 2017 for recent attempts to do this).

Second, in transport not only is the relationship between activities and mobility changing over time, so are the people undertaking these activities. As an example, Stapleton, Sorrell and Schwanen's (2016) analysis of rebound covers the period 1970 to 2011. In 1970, fewer than 35 per cent of 60- to 70-year-olds and 15 per cent of 70+-year olds had a driving licence. In 2011, this was 90 per cent and 79 per cent respectively. Changes in the composition of the driving population are relevant in that different age groups make different kinds of journeys to different sorts of destinations meaning that there are correspondingly different relationships with 'the car' over time. Separating out what is a result of 'the change to vehicle efficiency' from what is 'the change to mobility' requires substantial simplification of what underpins the average figures referred to above. There are also geographical considerations. Gillingham (2014) shows that the elasticity of demand for driving with respect to the price of gasoline is significantly different across different counties in California. The spatial scale of the analysis will therefore determine the size of the rebound effect, reinforcing its position as a social construct. Stapleton, Sorrell and Schwanen (2016) have used Great Britain as their unit of analysis but there will clearly be different effects occurring in London compared to those in rural Wales due to the types of journeys being undertaken and the options available.

While all of the literature reviewed here is admirably clear on the limitations of the approaches adopted, the limitations persist and could be substantial (Sorrell, 2007). Stapleton, Sorrell and Schwanen conclude, for example:

> [T]aking changes in fuel efficiency as the explanatory variable, we find little evidence of a long-run direct rebound effect in Great Britain over this period. However, taking changes in either the fuel cost of driving or fuel prices as the explanatory variable we estimate a direct rebound effect in the range 9% to 36%.
>
> *(Stapleton et al., 2016, p. 313)*

As Sorrell notes, 'the definition of inputs and outputs, the appropriate system boundaries for measures of energy efficiency and energy consumption and the timeframe under consideration can vary widely from one study to another' (Sorrell, 2007, p. 11). This undermines the comparability of different studies and also, to some degree, the notion of there being a strong science of rebound as each estimate is defined by the specificity of the study.

What are the implications of challenging the concept of rebound?

The purpose of research on rebound is to understand the likely impacts of new products, services, or policies on energy consumption. This aim seems entirely worthy of academic and policy endeavour given the discrepancy between what it is assumed 'efficiency policies' will achieve and what they actually do.

In developing scientific approaches to understand rebound, the key elements of change (linked to new technologies/services) have been identified. New products or services will change the way in which things are done. So, a new washing machine may contribute to an increase in the frequency of washing or the size of loads washed, offsetting some of the energy it is supposed to save. These direct changes attributable to specific products in use seem easiest to assess. Yet, there are also broader changes which can be created through product and service innovations which will play out elsewhere in the economy and which might have apparently little to do with washing machine use or vehicle efficiency or household insulation. These are considered but rarely captured in the current rebound literature. They may in fact be too difficult to trace as the world is not a controlled experiment and there are many changes constantly playing out over space and time.

Even so, it should be possible to think about how appropriate direct assessments of rebound effects are as proxies for total effects by considering different types of product or service change. For example, a change to cavity wall insulation may lead to a change in heating timings and temperature in the home. In this case 'rebound' questions focus exclusively on what the money saved in heating costs get spent on? However, innovations which free up time such as the first unsupervised washing machines or a fully autonomous vehicle would allow the patterning of activities to change significantly and, in the case of autonomous vehicles, the emergence of potential 'new demands' such as sending food to an elderly relative on the other side of town. The trickle down effects of some of these innovations are very difficult to estimate. Who would have foreseen the full extent to the automobile revolution when the Ford Model T launched in 1908?

The leading scholars in the field demonstrate and debate both the definitional, data and analytical challenges to generating convincing estimates of rebound. Despite the artificial positions and analytical stretching sometimes required to operationalise this concept, estimates of the first order impacts of energy innovations in use (which studies of the rebound effect help describe) are likely to be useful to policy. Analyses of rebound are much less capable of capturing the broader social transformations that innovations help create. Accordingly, the concept of rebound can only ever provide a partial view of the changing demand for energy. One aim in debating rebound in these terms is to show that the very idea demonstrates the need to think beyond the seemingly 'obvious' impacts of policies and technologies.

Notes

1 *Oxford English Dictionary*, s.v. 'rebound', https://en.oxforddictionaries.com/definition/rebound [accessed 28 February 2019].
2 Stapleton et al. (2017, p. 313) also note that 'for econometric studies, the most obvious measure of the direct rebound effect is the elasticity of demand for the relevant energy service (S) with respect to some measure of energy efficiency (ε): $\eta\varepsilon(S) = \partial\ln(S)/\partial\ln\varepsilon$. For example, the energy service provided by private cars may be measured in vehicle kilometres (vkm), their fuel consumption (E) in megajoules (MJ) and their fuel efficiency ($\varepsilon = S/E$) in km/MJ.'

Further reading

Berkhout, P. H. G., Muskens, J. C. and Velthuijsen, J. W. (2000). Defining the rebound effect. *Energy Policy*, 28(6–7), pp. 425–432.
Sorrell, S. (2007). *The rebound effect: An assessment of the evidence for economy-wide energy savings from improved energy efficiency*. London: UK Energy Research Centre. Retrieved from www.ukerc.ac.uk/programmes/technology-and-policy-assessment/the-rebound-effect-report.html [accessed 3 September 2018].

References

Gillingham, K. (2014). Identifying the elasticity of driving: Evidence from a gasoline price shock in California. *Regional Science and Urban Economics*, 47(C), pp. 13–24.
Gillingham, K., Rapson, D. and Wagner, G. (2016). The rebound effect and energy efficiency policy. *Review of Environmental Economics and Policy*, 10(1), pp. 68–88.
Jevons, W. S. (1866). *The coal question* (2nd edition). London: Macmillan and Company.
Llorca, M. and Jamasb, T. (2017). Energy efficiency and rebound effect in European road freight transport. *Transportation Research Part A*, 101, pp. 98–110.
Sorrell, S. (2007). *The rebound effect: An assessment of the evidence for economy-wide energy savings from improved energy efficiency*. London: UK Energy Research Centre.

Retrieved from www.ukerc.ac.uk/programmes/technology-and-policy-assessment/ the-rebound-effect-report.html [accessed 3 September 2018].

Sorrell, S., and Dimitropoulos, J. (2008). The rebound effect: Microeconomic definitions, limitations and extensions. *Ecological Economics*, 65(3) pp. 636–649.

Stapleton, L., Sorrell, S. and Schwanen, T. (2016). Estimating direct rebound effects for personal automotive travel in Great Britain. *Energy Economics*, 54, pp. 313–325.

Stapleton, L., Sorrell, S. and Schwanen, T. (2017). Peak car and increasing rebound: A closer look at car travel trends in Great Britain. *Transportation Research Part D*, 53, pp. 217–233.

Vivanco, D. F., Kemp, R. and van der Voet, E. (2016). How to deal with the rebound effect? A policy-oriented approach. *Energy Policy*, 94, pp. 114–125.

Walnum, H., Aall, C. and Løkke, S. (2014). Can rebound effects explain why sustainable mobility has not been achieved? Sustainability, 6(12), pp. 9510–9537.

6

ELASTICITY

Jacopo Torriti

Spiders make incredibly elastic webs capable of withstanding all sorts of stresses and strains. The concept of price 'elasticity' is different. It refers to peoples' willingness to forego services if prices rise. Activities that have to be done at almost any cost are sticky and inelastic. Others are more flexible. This chapter is about the rigidities and springyness in daily life and how these are organised.

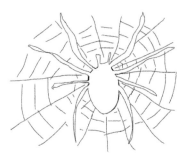

Introduction

Elasticity of demand is generally understood in (energy) economics as a way to measure the relationship between price and quantity (of energy) demanded in a given time period. In essence, elasticity is a direct measurement of how increases in the price of the supplied good or service correspond to decreases in the quantity of demand. Measuring variations in energy demand over time due to price changes is equivalent to saying that there are substitutes for energy. For instance, if a certain brand of pasta becomes very expensive, one could simply switch to another brand. Demand for one brand of pasta or another is highly elastic. The same cannot be said about energy demand.

Measurements of elasticity vary over different temporal scales. Long-term elasticity is generally across a decade or more, short term is over a year and extremely short is within the day. Short-term elasticity is particularly important for those interested in understanding how effective demand-side interventions to modify electricity demand based on price might be.

Conventional wisdom about elasticity holds that changes in energy demand over time are in proportion to changes in price and that price is the one thing that matters for understanding how much energy is used. Alternative narratives start from the view that there are many more variables that account for changes in energy demand over time.

The origins of elasticity

The concept of elasticity originates from neo-classical approaches in micro-economics which sought to measure and quantify the relationship between price and demand. The modern history of applying these ideas to energy demand dates back to the 1970s oil crises, which triggered attempts to reduce growing demand for electricity, and to stem increasing dependency on oil imports and negative environmental impacts. It might sound odd, but the high costs of producing electricity in the late 1970s meant that several utilities started developing programmes designed to *reduce* the final price of electricity in order to ensure that demand did not collapse. This is a result of one of the most extensively explored theories on the direct effects of oil price shocks (i.e. input-cost effect). According to this theory, higher energy costs lower the use of oil, which in turn lowers the productivity of capital and labour.

The input-cost effect theory was accompanied by an income effect theory, stating that the higher cost of imported oil reduced households' disposable income. The economic context of that historical period, combined with the regulatory framework compelled utilities to actively limit the final price of electricity to prevent 'elastic' effects of the kind that had already been experienced by the automobile industry. During the oil crisis the automobile industry was devastated by the collapse of consumer demand for large and inefficient cars (Lee and Ni, 2002). This is an example of high elasticity where the change in demand is greater than the change in price (interest in owning such a car fell faster than the cost rose). In practice, the extent to which an even more significant increase in the price of electricity in the late 1970s would have triggered a collapse in demand is still debated.

In studies of energy markets, elasticity is concerned with the basic issue of allocating scarce resources in the economy. This ranges from microeconomic concerns about energy supply and demand through to macro-economic concerns about investment, financing and economic linkages with other markets. At the same time, the issues facing the energy industry change and the role of elasticity evolves accordingly.

In its first formulation, price elasticity was seen as complementary to energy substitution, because if elasticity is high so is the potential for substitution. In the case of pasta brands, elasticity is high, which means that one brand can be easily substituted by another. Over time, the price elasticity of electricity demand emerged as a way of measuring how trends in electricity prices could affect demand (i.e. long-term elasticity) and, even more recently, how variations in tariffs at different times of the day could affect demand (i.e. short term elasticity).

In the energy economics literature, the evidence on which studies of elasticity draw varies. An old meta-analysis by the World Bank (1981) summarises the problematic evidence around price elasticity of energy demand by saying that it is 'uncomforting to find out that the statistical and other estimates of energy demand elasticities are widely divergent' (World Bank, 1981, p. 7).

A typical example of an empirical study in this area is that by Silk and Joutz (1997), who find that a 1 per cent increase in electricity prices reduces electricity consumption by 0.62 per cent. Other more recent studies show that the elasticity of electricity demand ranges from very low (Blázquez, Boogen and Filippini, 2013) to very high (Krishnamurthy and Kriström, 2013). But in general, it is understood that the elasticity of electricity demand is low. For example, for intra-day price elasticity (or extreme short run price elasticity) there seems to be evidence that parents do not postpone by an hour the moment when they drop children off at school to take advantage of a profitable off-peak tariff. In other words, the price of energy is not a major driver of people's everyday lives.

Elasticity in energy policy and in the market

For policy-makers, the assumptions underpinning a linear relation between price and demand are at the heart of much of the reform in electricity markets. Four recent examples are outlined here – two from policy and two from market applications. First, the goal of increasing the elasticity of demand was one of the main drivers for reforming retail markets and increasing competition. In assessing the policy impact of restructuring electricity retail markets in US states in order to encourage greater 'demand side'/consumer participation, Su (2015) finds that 'only residential customers have benefitted from significantly lower prices but not commercial or industrial customers. Furthermore, this benefit is transitory and disappears in the long run.' As this example shows, policies driven by the belief that elasticity will deliver changes and reductions in demand have not yielded the expected returns.

Second, in the UK Ofgem has recently been considering the introduction of dynamic tariffs, following recommendations by the Competition and Markets Authority (2016). Dynamic tariffs are designed to be cost-reflective and include time of use tariffs, which encourage consumers to shift consumption away from periods when demand on the network is higher (peak periods) or reduce it altogether. In other words, the idea is that dynamic pricing should increase the flexibility on the demand side thanks to consumers' exposure to varying electricity prices. This supposes, perhaps erroneously, that households are willing and able to respond to price signals by altering their energy consumption.

Third, as in other markets where sudden increases in demand push prices up, energy suppliers have been making money at times of peak demand thanks to higher marginal prices. The reason why this was allowed to happen relates to the combination of a flat energy price and the relatively low elasticity of demand. At times of peak demand, increasing the output of gas-fired generators makes wholesale electricity prices more volatile. Electricity price spikes occur frequently when demand is very high because there is extremely high price volatility due to the higher volume risks. At peak time less efficient or higher cost generating units are used to meet higher demand. In such circumstances, electricity prices in wholesale markets can fluctuate significantly on several days of the year.

Fourth, a logical step which follows on from price elastic demand is substitution. If prices for a fuel go up significantly, substitute fuels will be sought after. This principle is operationalised via two types of substitution. First, interfuel substitution is associated with situations in which each utility decides on a combination of fuel use based on the availability of their generation units, previous investment decisions, regulatory constraints and the price or costs of fuels (Moxnes, 1990). Second, cross price elasticity of demand measures the responsiveness in the quantity of demand for one fuel when a change in price takes place in another (e.g. gas) (Kirschen et al., 2000). The measurement is calculated by taking the percentage change in the quantity demanded of one good and dividing it by the percentage change in price of the other good (as in the example of the pasta at the beginning of this chapter). Cross price substitution has to some extent been possible in transport but much less for energy in buildings. Cars are purchased with a higher frequency than homes. In addition, the infrastructure in and around the home makes it difficult to switch between gas and electricity.

In summary, the concept of elasticity has proved critical in motivating major changes in energy policy and markets: it sustains the introduction of reforms aimed at increasing demand side participation; the introduction of dynamic, or time of use tariffs and fuel substitution.

Problematic assumptions about elasticity

Underpinning the concept of elasticity is the idea that price is the main factor influencing changes in energy demand over time. This idea is a consequence of plotting demand curves as functions of price and quantity. Economists who have criticised the concept of price elasticity in relation to commodities (e.g. Samuelson, 1965) have emphasised how arbitrary it is because of subjective judgements made about values that pertain in the first time period in relation to which elasticity is measured, and because the prices of all other commodities are assumed to be constant.

Like other economic measurements, the relationship between price and volumes of energy demand is supposed to exist *ceteris paribus*. For decades economists have been measuring the price elasticity of energy demand notwithstanding changes in almost everything else, including: temperatures, intensities of different fuels per units of input and, arguably most important of all, the purposes of energy provision. As a result, the exercise of calculating price elasticities across different temporalities is conceptually and methodologically flawed.

Treating energy demand as if it was the same as demand for any other good or service, whose quantity varies depending on modifications in price, depends on ignoring what energy is for. Energy enables people to meet other demands, for instance, for cleaning (bodies, clothes and dishes), cooking, lighting, heating, working and entertainment. To assume that these practices will vary in intensity because of variations in short term prices or energy bills is equivalent to disregarding the realities of daily life.

Fundamentally, what energy economists measure is an average elasticity, which does not reflect the temporalities of what people do at different times of the day. Even following the rationale of short-term price elasticity, it seems counterintuitive to suppose that people have the same opportunities to respond to price changes that occur at different times of the day. A further complication is that the figures do not stack up as expected. For instance, when looking at the yearly aggregate domestic electricity demand in the UK between 2005 and 2011 (see Figure 6.1), there is a clear decrease, despite the fact that increases in prices were not dissimilar from those experienced in the preceding decade. This suggests that the intensity of demand can reduce *per capita* irrespective of price.

The 'fable' or recurrent discourse of price elasticity can be challenged empirically and conceptually. Conceptually, two simple points subvert the basic idea that energy demand depends on price:

1.   what people do changes over time and energy demand follows suit; and
2.   changes in demand affect prices.

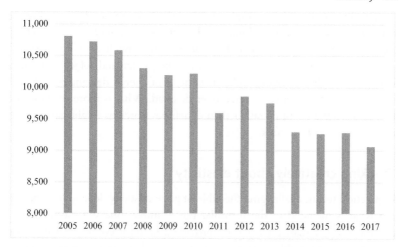

FIGURE 6.1 Domestic electricity consumption (tonnes of oil equivalent) by year (2005–2017) in the UK.

Source: based on data from DUKES 2018 (Department for Business, Energy & Industrial Strategy, 2018)

The first point is not intuitive for economists as it implies that practices are the main independent variable driving demand. People face capital costs associated with the purchase of appliances, devices, lamps, cars, etc. When these items are bought they are subject to known price-quantity relations of demand. However, electricity, gas and water are bought at a price, but not one that relates to the depreciation of capital cost.

The second point goes back to the relationship between price and quantity. The suggestion that demand affects prices might have been counterintuitive in a twentieth-century economy in which the costs of coal, oil and gas figure as defining factors for generating and supplying energy. However, it becomes progressively more relevant in a twenty-first-century low carbon economy which increasingly relies on renewable sources of energy supply, in which fuel costs next to nothing and in which prices are more obviously dictated by changes in demand.

An example of this comes from Germany. On Sunday 16 June 2013, between 2.00 and 3.00 p.m., electricity demand was low. Electricity supply was very high because it was a sunny and windy day. As a result, the wholesale price of electricity dropped to a negative 100 euros per MWh to protect the grid from overloading. In effect, generating companies were forced to pay the system operator to take their electricity. This is a great example of how demand can drive changes in price.

The two points above trigger further thinking about the meaning of elasticity outside the price domain. Elasticity is the ability of an object or material to resume its normal shape after being stretched or compressed. As is now obvious, the stretching of the (energy demand) object does not happen through price alone. The inelasticity of energy demand reveals the relative lack of influence of price over demand. What is more, it indicates that by focusing on price rather than demand researchers might get hold of the wrong end of the stick.

Thinking creatively about elasticity

In practical terms, challenging the fable of price elasticity depends on changing what is included in the formula. For example, one suggestion is that a variable other than price would provide a better account of changes in demand. Three possible variables are put forward here: (i) occupancy elasticity; (ii) heating and occupancy-based elasticity; and (iii) time-only elasticity.

Occupancy elasticity is defined and measured as variation in the quantity of energy demand in a household, given changes in occupancy over time. Occupancy elasticity starts from inelasticity (for unoccupied households) and increases with occupancy (e.g. when more people enter the household). For instance, at peak time, where energy demand increases more than occupancy rates, there might be high inelasticity of energy demand.

Taking as a starting point the suggestion that heating varies according to outdoor temperature and occupancy, the elasticity of heating demand could be based on changes in temperatures and occupancy (rather than price). A heating and occupancy-based measure of elasticity would determine: (i) whether changes in temperatures have a higher weight than occupancy as regards changes in heating demand or (ii) whether changes in occupancy have a higher weight than temperatures as regards changes in heating demand. A time–only version of elasticity could simply describe how the activities that people perform change over different periods of clock-time, (e.g. from one 10-minute period to the next) and how these changes affect energy demand.

There are other routes to take. Alternative analyses of price elasticity provide some clues as to the direction that unconventional approaches to temporality and energy demand might follow.

On non-averaged price elasticity: a recent study by a group of Swedish researchers looks at elasticity changes during the day for lighting and cooking and to a lesser extent for heating (Broberg and Persson, 2016). This involves a move away from average elasticities an attempt to incorporate the time of the day and aspects of space within the household into price elasticity calculations.

On occupancy elasticity: Torriti (2012) looks at occupancy levels of single-person households in 15 different European countries. This work measures how occupancy varies within 10-minute intervals based on time use data. Changes in occupancy throughout the day vary significantly across countries. For instance, in southern European countries occupancy is generally higher later in the evening and, correspondingly, peaks in residential electricity demand take place later in the day.

On time-only elasticity: some research in this direction was carried out as part of quantitatively estimating the clock-time dependence of household practices (Torriti, 2017). An analysis of UK time use data shows that washing has the highest value on the time dependence metric; using computers is the least time-dependent practice; Tuesdays, Wednesdays and Thursdays have the highest time dependence for all practices; and certain energy-related practices have higher seasonal dependence than others.

These examples are starting points and all need more conceptual work and empirical research. Once we understand the marginal role of price of energy in everyday lives, and once it becomes clear that the price elasticity of energy demand is a mythical concept, it will be possible to focus on developing more realistic interpretations of the dynamics of energy demand.

Further reading

Broberg, T. and Persson, L. (2016). Is our everyday comfort for sale? Preferences for demand management on the electricity market. *Energy Economics*, 54, pp. 24–32.

Torriti, J. (2012). Demand side management for the European Supergrid: Occupancy variances of European single-person households. *Energy Policy*, 44, pp. 199–206.

Torriti, J. (2017). Understanding the timing of energy demand through time use data: Time of the day dependence of social practices. *Energy Research and Social Science*, 25, pp. 37–47.

References

Blázquez, L., Boogen, N. and Filippini, M. (2013). Residential electricity demand in Spain: New empirical evidence using aggregate data. *Energy Economics*, 36, pp. 648–657.

Broberg, T. and Persson, L. (2016). Is our everyday comfort for sale? Preferences for demand management on the electricity market. *Energy Economics*, 54, pp. 24–32.

Competition and Markets Authority (2016). *Provisional decision on remedies*. London: Competition and Markets Authority. Retrieved from https://assets.publishing. service.gov.uk/media/56efe79040f0b60385000016/EMI_provisional_decision_ on_remedies.pdf [accessed 23 November 2018].

Department for Business, Energy & Industrial Strategy (2018). *Digest of UK energy statistics 2018*. London: Department for Business, Energy & Industrial Strategy. Retrieved from www.gov.uk/government/collections/digest-of-uk-energy-statistics-dukes#2018 [accessed 15 February 2018].

Kirschen, D. S., Strbac, G., Cumperayot, P. and de Paiva Mendes, D. (2000). Factoring the elasticity of demand in electricity prices. *IEEE Transactions on Power Systems*, 15(2), pp. 612–617.

Krishnamurthy, C. K. and Kriström, B. (2013). *Energy demand and income elasticity: A crosscountry analysis.* Working Paper 2013: 5. Umeå: Centre for Environmental and Resource Economics.

Lee, K. and Ni, S. (2002). On the dynamic effects of oil price shocks: a study using industry level data. *Journal of Monetary Economics*, 49(4), pp. 823–852.

Moxnes, E. (1990). Interfuel substitution in OECD-European electricity production. *System Dynamics Review*, 6(1), pp. 44–65.

Samuelson, P. A. (1965). Using full duality to show that simultaneously additive direct and indirect utilities implies unitary price elasticity of demand. *Econometrica: Journal of the Econometric Society*, 33, pp. 781–796.

Silk, J. I. and Joutz, F. L. (1997). Short and long-run elasticities in US residential electricity demand: A co-integration approach. *Energy Economics*, 19(4), pp. 493–513.

Su, X. (2015). Have customers benefited from electricity retail competition? *Journal of Regulatory Economics*, 47(2), pp. 146–182.

Torriti, J. (2012). Demand side management for the European Supergrid: Occupancy variances of European single-person households. *Energy Policy*, 44, pp. 199–206.

Torriti, J. (2017). Understanding the timing of energy demand through time use data: Time of the day dependence of social practices. *Energy Research and Social Science*, 25, pp. 37–47.

World Bank (1981). *Energy demand elasticities: Concept evidence and implications.* Washington, DC: World Bank.

PART III
Injunctions

7

LOW HANGING FRUIT

Elizabeth Shove and Noel Cass

Low hanging fruit are easy to pick. There is no need to climb the tree, or reach for a ladder. In the energy sector, the phrase 'first pick the low hanging fruit' is an instruction to start by harvesting the 'easy wins'. This chapter discusses the trees and fruits of this metaphorical orchard.

Introduction

According to the *Collins English Dictionary*, the phrase 'low hanging fruit' refers to (1) the fruit that grows low on a tree and is therefore easy to reach and (2) a course of action that can be undertaken quickly and easily as part of a wider range of changes or solutions to a problem.[1] In the energy field, picking the 'low hanging fruit', means first taking measures that are simple to implement and that generate quick returns in terms of carbon and energy demand reduction. There are different interpretations of what count as 'low hanging fruit', but Kevin Anderson's list gives a sense of the kinds of actions that fall into this category:

- All new cars sold in the UK should meet a minimum mpg standard by 2010.
- Fridges and freezers sold after 2008 should not exceed a maximum energy use.

- New (best practice) building regulations should be raised incrementally at 2 year intervals.
- Phase out standby facilities on electrical gadgets, or insist on tough consumption standard.
- Phase out all 'normal' light bulbs by 2008 to be replaced by low-energy bulbs.
- Growth in aviation should be limited; expansion at existing airports should curtailed.

(Anderson, 2006)

Companies selling energy services and advice talk of 'low hanging fruit' as a means of underlining the benefits of what they do. For example, one such organisation claims that in the US: '160 Megatons of global-warming CO_2 that are being emitted [by households] . . . can be eliminated by following some simple steps!' (Ekotrope, undated). Another concludes that 'picking these low hanging fruit, can easily uncover 10–15% [energy savings] or more in no cost and low cost changes alone' (Schwartz, 2014). The concept is also used by environmental lobby groups: 'Household energy consumption could be reduced by about 22 percent (8 quads) by behavioral improvements such as keeping tires properly inflated and placing refrigerators a few inches away from the wall' (EESI, 2009). And within areas of energy and environmental policy, efficiency is said to be one of the lowest hanging fruit on the tree (Johnston, 2017).

The notion of 'low hanging fruit' brings together a series of related ideas. One is the self-evident logic of taking simple steps first. Stephen Chu, at the time US Secretary of Energy, made this plain in saying that 'energy efficiency is not just low-hanging fruit; it is fruit that is lying on the ground' (Chu, 2009). The phrase also supposes that forms of intervention (here energy demand reduction) can be positioned on a spectrum of difficulty ranging from measures that are 'easy' to implement, at one end of the scale, through to others that are harder, meaning that they are more costly, more contentious or more complex. As discussed below, the metaphor of 'low hanging fruit' reproduces other assumptions about energy demand and the actions and policies that can be taken to reduce consumption. In taking this term seriously, and in taking it apart – in asking 'What are the fruits?', 'What kind of tree do they grow on' and 'What does picking involve?' – we show how this language situates and frames both the problem and the politics of energy demand reduction.

This is not a new topic and as discussed below, there are established critiques of the short-termism of plucking all the 'low hanging fruit', and discussions of the risks and dangers that follow (Robèrt, 2000; Narain and Van't Veld, 2008). Such criticisms point to important questions about the size of fruit (where are the real gains to be had?), and about the politics of harvesting. Other also critical analyses call into question the very foundations of the metaphor, arguing that it reproduces and reinforces a distinctly technocratic approach that bypasses fundamental questions about what energy is used for in society, and how this changes over time. On both counts, it seems that a more sophisticated conceptualisation of energy demand calls for a new lexicon, and new set of images and ideas. Before thinking about what this might entail it is useful to consider the characteristics of the three main entities involved in this popular metaphor, starting with the fruit.

The fruit

The metaphorical fruit that grow on the imaginary tree of energy demand reduction have a number of distinctive properties. First, they are easy to identify and separate. They can be picked off, like apples, one by one. Second, individual fruits do not re-grow: they can only be harvested once and when they are gone, they are gone. Third, each fruit has a definite position on the tree: each is fixed at a certain height above the ground. These features are entirely consistent with the ways in which energy efficiency measures are conceptualised in research and policy. For example, replacing an old boiler is a discrete and bounded action, akin to picking an apple. Swapping inefficient light bulbs represents another, also well-defined action. And in both cases, once such actions have been taken, and when the new more efficient boilers or light bulbs are in place, there are no further gains to be had (at least not for a while). In areas like car engine design, some go so far as to conclude that 'We've picked all the low-hanging fruit when it comes to fuel efficiency' (Wald, 1990), implying that are only so many 'easy' opportunities that can be taken.

There are some known complications with these ideas and with the logic on which they depend. For example, when all the low hanging fruit have been picked, higher fruits that were initially harder to reach become the lowest available. In addition, the exact positioning of fruits on the tree can change over time. If certain technologies become cheaper, the fruits associated with them become easier to reach. And if new more efficient technologies are developed, new fruits appear. As David Goldstein of the US natural resource defence council explains, 'the tree that grows this fruit is special as well: when you pick the fruit, as we have almost a dozen times with [efficiency standards for] refrigerators, it grows back' (Goldstein, 2011). Where this kind of re-growth occurs, it is usually gradual.

These developments aside, at any one moment, fruits on the upper branches require greater effort to pick than those closer to the ground. Given this vertical distribution of cost and benefit, stripping the tree of all fruits within easy reach significantly reduces the relative value of the remaining crop. This leads some energy economists to argue for a mixed strategy in which low and high hanging fruits are combined in the same basket, and in which enough financially 'juicy', low hanging options are left to make picking an attractive proposition, even in the longer run (ThinkProgress, 2012).

In some cases, it is easy to rank energy saving options in these terms. However, judgements about the balance between effort and potential return are often complicated by the fact that many forms of demand reduction have effect in combination meaning that their value cannot be determined in isolation. For example, measures that count as 'low hanging fruit', like installing a more efficient boiler, may be even more effective if combined with additional loft or wall insulation. Does this mean that some fruits hang lower if they are picked together (Adams, 2015)? Answers to this biologically puzzling question depend, in part, on the properties of the tree.

The tree

In most representations, the imagined tree is an essentially static framework from which fruits dangle. Its height and the distribution of its branches are important, but the metaphor of 'low hanging fruit' takes no account of how high the tree grows or of how fruits develop and form. There is no reference to pruning, cultivating or fertilising; to the environment in which the tree is situated, or to seasonal change. This characterisation of the tree is consistent with representations of technological efficiency as an absolute rather than a relative achievement, and as something that is unaffected by surrounding and constantly changing societal landscapes of activity and demand.

The image of a single tree with a limited crop of fruits has other connotations. As mentioned above, some fruits might re-grow (as new technologies are introduced), but the size of the tree, and by implication the size of the problem is itself unchanging. All other things being equal it is in theory possible to imagine a time when even the highest fruits have been harvested, and when there is no more to do. In practice, energy researchers and policy makers rarely subscribe to such a finite view of the future. However, many do adopt the working hypothesis that patterns of demand will remain much as at present and that their task is to meet these needs as efficiently as possible. This is how many pickers define their role.

The fruit pickers

Although policy makers and researchers are key, consumers and manufacturers are also involved in picking the low and high hanging fruit of energy demand reduction. This mixing of actors is sometimes problematic. It is so in that different types of pickers almost certainly encounter different types of fruit/opportunity. For example, the actions that householders, manufacturers, or policy makers can take to reduce energy demand, and what counts as a low or a high hanging fruit from each of these different points of view are not the same. In the realm of policy making, 'easy' wins are typically associated with the use of established policy measures (pricing, regulation), with 'harder' prizes being those that require shifts in culture or in consumer response, or that depend on joined up policy across multiple sectors. By contrast, within households, easy actions are generally thought to be those that are consistent with current practice, or that have few consequences in terms of cost and time.

Rather than articulating these differences, the language of low hanging fruit emphasises two seemingly common features. One core assumption is that members of all these different 'picking communities' have the capacity to act: in other words, each is able to reach up, grab a fruit and reap the benefit. A second uniting factor is that the basis for selecting one fruit rather than another is essentially the same. In all cases, picking strategies are assumed to depend on bounded (individualised) forms of cost-benefit analysis in which the efforts of climbing and harvesting are weighed against the gains that follow. One further feature is that imagined pickers implicitly act alone, and of their own volition. In other words, there is no collusion in the orchard, no jumping up to bend the branches down for someone else, and no conscripted labour either.

The concept of picking the 'low hanging fruit' is a common theme in journalistic and business-oriented discussions of energy reduction and climate change, but how are these ideas mobilised in policy making and research? As outlined in the next two sections, fruit picking is not as simple as it seems.

The politics of the orchard

If there are significant benefits to be had from loft insulation, or from increasing the efficiency of aircraft engines, why not take advantage of these opportunities? At first sight, the focus on easy pickings seems to make sense. However, various commentators argue that such strategies are at best relatively ineffective, and at worst counterproductive. For example, Jeffrey Sachs, an economist and adviser to the UN, distinguishes between forms of 'deep decarbonisation' that pave the way for still more radical forms of carbon reduction, and responses that generate quick but comparatively modest results (picking low hanging fruit). In brief, the argument is that picking 'low hanging fruit' is a strategy that offers quick wins but that also reproduces energy systems and ways of life that are, in fact, part of the problem. According to Sachs, picking the fruits of efficiency does not help societies to escape the high carbon 'trap' (SDSN and IDDRI, 2014), nor does it actively promote or engender much lower carbon ways of living.

This critique argues for a new approach. In essence, it suggests that policy makers need to pay much greater attention to the potentially very large fruits situated high up in the crown of the tree. The basic idea is that this is the level at which the transition to a low carbon society has to occur. As well as calling for nifty ladder work, such an analysis hints at a more complex relationship between the size and location of the imagined fruit and the effort involved in picking them.

The potential conflict between short-term opportunities and longer-term goals points to an important distinction between efficiency on the one hand and the much broader ambition of demand and carbon reduction on the other. To illustrate, in the transport sector there is evidently scope for increasing the efficiency of car engines and transmission systems. However, carbon emissions could be reduced much more dramatically if supply chains were shortened, if production was localised and if systems of urban planning and city management enabled people to access work, health and other necessary services by walking, cycling or using public transport (Banister, 2008). Rather than starting from the bottom and working upwards, the rationale of first picking the low hanging fruit is reversed. In this case, starting from the top depends on identifying opportunities for systemically reducing demand and for doing so on a societal scale (why are people travelling, where are they going) and only then turning to the 'easier', but quantitatively less significant challenge of making more efficient cars.

These are not exclusive strategies, but they depend on very different interpretations of the format of the tree, and of the characteristics and the distribution of larger and smaller fruits. They also reflect different interpretations not only of picking, but of intervention in general. To stretch the metaphor to its limit, those whose aim is to harvest the fruits of efficiency have almost no interest in the tree. By contrast, those who start from the top aim to modify the structure of demand, as such.

From fruit trees to fungi

Whether the strategy is to first pick the low hanging fruit, or to climb higher and start at the top, dominant discourses treat 'energy' as a finite resource that can be organised and managed. It is this basic understanding that makes it possible to talk of larger and smaller returns, and to distinguish between lower and higher hanging fruit. In addition, and as mentioned above, this terminology supposes that decisions about which energy saving measures to adopt (what to pick, and when) are made by autonomous and variously capable agents, whether these be policy makers, manufacturers or consumers. From this point of view, the image of a person reaching out to select a fruit accurately represents the conceptualisation of human agency as something that is distinct and separate from the realm of technology.

Despite being powerful and pervasive, these representations are at odds with much of what we know about the dynamics of energy demand, and about how technologies (including efficient ones) constitute the social practices on which resource consumption depends (Shove and Walker, 2014). If patterns of consumption are, to an extent, defined by the technologies involved, and if material elements (resources, devices, infrastructures) are constitutive of social practice (Shove, 2017), the defining features of the 'fruit' metaphor do not ring true.

Rather than thinking of fruit, picker and tree as separate entities, energy demand (overall) depends, at a minimum, on how these analytic categories are defined and conjoined, and how forms of connectivity develop and evolve over time. For example, over the last century or so, many activities that previously depended on human labour have been 'delegated' to machines of one kind or another. In aggregate, such processes generate the need for networks and infrastructures of power; they help constitute categories like those of consumer and provider, and they simultaneously transform the world of goods and the substance and character of everyday life.

Since images and metaphors structure ways of knowing and acting in research and in policy, and since the language of fruit trees does not capture these crucial aspects of emergence, horizontal development and co-constitution, new figures of speech (and thought) are required. There are many ways to go, but one option is to turn to social theory, and to draw, very selectively, from the writings of Deleuze (1988). Deleuze discusses the qualities of knowledge, emphasising forms of connection, intersection and circulation and using the biological metaphor of a rhizome to signify the complex webs, nodes and networks involved.

This way of thinking about knowledge is contrasted with other more hierarchical, 'arborescent', or tree-like representations, organised around clear-cut categories and binary distinctions. These ideas were certainly not developed with energy policy in mind, but the rhizome-metaphor captures some (though clearly not all) of the processes through which energy systems and related systems of consumption and practice evolve.

For instance, rather than imagining vertical structures of trees, fruits and ladders, we could represent the technologies and practices of energy demand (and demand reduction) as rhizomatic arrangements that spread and change and interact, absorbing 'nutrients' and transforming while also being transformed by their surroundings. This more fungus like imagery reflects and enables a different way of conceptualising the constitution and the dynamics of energy demanding practices.

This is not simply a matter of acknowledging 'complexity' or of noticing processes of grafting, nourishing and seeding in addition to those of picking. Instead, the more challenging suggestion is that policy makers, researchers and consumers are part of always shifting geometries and configurations of people, resources and things. Terms and languages that foreground these interconnections do not to deny the scope for policy intervention or for effective action. However, they do imply that this is more a matter of modulating pathways and trajectories than of simply harvesting gains in efficiency. From this point of view, the image of mycelium threading through while also being part of an always emergent living system makes for a less picturesque but more accurate representation of energy policy in practice than the image of fruits hanging from a tree.

Note

1 *Collins English Dictionary*, s.v. 'low hanging fruit', www.collinsdictionary.com/dictionary/english/low-hanging-fruit [accessed 20 November 2018].

Further reading

Shove, E. and Trentmann, F. (eds) (2019). *Infrastructures in practice: The dynamics of demand in networked societies*. London: Routledge.

References

Adams, N. (2015). How 'low-hanging fruit' has poisoned energy efficiency. Retrieved from www.greentechmedia.com/articles/read/why-low-hanging-fruit-has-poisoned-energy-efficiency [accessed 20 November 2018].

Anderson, K. (2006). Energy's 'low hanging fruit'. Retrieved from http://news.bbc.co.uk/1/hi/sci/tech/4633160.stm [accessed 28 May 2018].

Banister, D. (2008). The sustainable mobility paradigm. *Transport Policy*, 15, pp. 73–80.

Chu, S. (2009) Secretary Chu opinion piece in *Times of London*. Retrieved from https://energy.gov/articles/secretary-chu-opinion-piece-times-london [accessed 28 May 2018].

Deleuze, G. (1988). *A thousand plateaus: Capitalism and schizophrenia*. Electronic resource. London: Continuum.

EESI (2009). Low-hanging fruit: The economics of energy efficiency. Retrieved from www.eesi.org/briefings/view/low-hanging-fruit-the-economics-of-energy-efficiency [accessed 20 November 2018].

Ekotrope (undated). Picking the low-hanging fruit: Improving the energy-efficiency of existing houses. Retrieved from https://ekotrope.com/picking-the-low-hanging-fruit-improving-the-energy-efficiency-of-existing-houses [accessed 20 November 2018].

Goldstein, D. B. (2011). The low-hanging fruit . . . that keeps growing back. Retrieved from www.nrdc.org/experts/david-b-goldstein/low-hanging-fruit-keeps-growing-back [accessed 20 November 2018].

Johnston, I. (2017). Environmental policies are driving down consumer energy bills, experts say. *The Independent*, 16 March. Retrieved from www.independent.co.uk/environment/environmental-policies-low-carbon-economy-renewable-cutting-energy-bills-committee-on-climate-change-a7631961.html [accessed 20 November 2018].

Narain, U. and Van't Veld, K. (2008). The clean development mechanism's low-hanging fruit problem: When might it arise, and how might it be solved? *Environmental and Resource Economics*, 40(3), pp. 445–465.

Robèrt, K.-H. (2000). Tools and concepts for sustainable development: How do they relate to a general framework for sustainable development, and to each other? *Journal of Cleaner Production*, 8(3), pp. 243–254.

Schwartz, M. (2014). Where is your building's 'low hanging fruit' for energy efficiency? Retrieved from www.alturaassociates.com/building-energy-efficiency-where-is-your-low-hanging-fruit [accessed 20 November 2018].

SDSN and IDDRI (2014). Pathways to deep decarbonisation. Retrieved from http://unsdsn.org/wp-content/uploads/2014/09/DDPP_Digit.pdf [accessed 20 November 2018].

Shove, E. (2017). Matters of practice. In A. Hui, T. Schatzki and E. Shove (eds), *The nexus of practices: Connections, constellations, practitioners*. London: Routledge, pp. 155–168.

Shove, E. and Walker, G. (2014). What is energy for? Social practice and energy demand. *Theory Culture and Society*, 31(5), pp. 41–58.

ThinkProgress (2012). Rotten fruit: Why 'picking low-hanging fruit' hurts efficiency and how to fix the problem. Retrieved from http://thinkprogress.org/climate/2012/11/23/1230521/rotten-fruit-why-picking-low-hanging-fruit-hurts-efficiency-and-how-to-fix-the-problem [accessed 20 November 2018].

Wald, M. L. (1990). In cars, muscle vs. mileage. *The New York Times*, 16 August. Retrieved from www.nytimes.com/1990/08/16/business/in-cars-muscle-vs-mileage.html?pagewanted=all [accessed 20 November 2018].

8

KEEPING THE LIGHTS ON

Gordon Walker

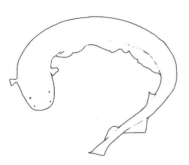

Electric eels have muscle-like cells, called electrocytes. These enable them to deliver electric shocks at any time of the day or night. Electric eels do not light up, but they can produce enough electricity to illuminate up to 20 light bulbs, for an instant (Tennessee Aquarium, 2015).

Introduction

Keeping the lights on is a recurrent phrase used in political and public discussion of energy issues. It conveys the idea that it is crucial to keep energy systems working, smoothly and uninterrupted, and for governments to ensure that this is achieved. While evidently an important goal, its commonplace use is grounded in a set of assumptions about how 'keeping the lights on' is to be ensured and how the relationship between supply and demand is to be managed.

Classic 'keeping the lights on' thinking sees energy demand as non-negotiable, lying outside the frame of legitimate policy debate. As such, it justifies a focus on energy resources and generation with the aim of ensuring sufficient supply to meet whatever demand exists at any point in time, with the implication that failure would be catastrophic: the lights would go out and chaos would follow. This thinking can be challenged. Keeping a wide range of energy flows and services going does not have to be just about endlessly focusing on and increasing supply. We are already starting to see responsive demand management as a way of reducing demand at peak times rather than meeting it with peak supply. But given the challenge of meeting CO_2 emission targets, a far more fundamental debate is also needed about how much energy is enough; or exactly how many lights (and other energy uses) need to be kept on, now and in a necessarily low carbon future.

What has 'keeping the lights on' come to mean?

There is evidently a direct literal meaning to keeping the lights on, that of sustaining electrically powered illumination and not letting 'the lights' go out. But the phase is also idiomatic, conveying more than just its literal meaning; or metaphorical, a figure of speech which is symbolic of a wider concern.

Examples abound in the media, in government statements and reports, in political rhetoric, and in marketing by companies working in the energy sector. An internet search finds examples related to energy issues in Latin America, Asia, Western and Eastern Europe and the US, making it a globally circulating idiom.[1] In the UK it tends to feature on a seasonal cycle as winter approaches and concerns are voiced by politicians and media commentators about how close we might come to electricity supply not being able to meet peak winter demand (see for example Guardian, 2016). But it is also mobilised by policy makers, industry lobbyists and non-governmental groups in relation to longer running debates about new supply-side decisions or investments, as well as energy company regulation (see for example Ratcliffe, 2016). It is very much, therefore, a phrase that has become commonplace, embedded in ways of thinking through its repetition and largely used without critique.

Captured in how the phrase is used is an apparently simple and powerful metric of success and failure. The lights not being kept on is symbolic of failure, of a system that is not fit for purpose and of a government that has not properly done its job in looking after a core need of the economy and its citizens. Keeping the lights on is to be 'secure and stable', to have resilience and to be successfully planning ahead to deal with the stresses and strains that an energy system is foreseeably likely to encounter, including predicted future patterns of demand, old power stations being decommissioned and established energy resources 'running out' or becoming more expensive. During the UK coal miners' strikes of the 1980s, Mrs Thatcher reportedly used the justification of 'keeping the lights on' (Marsden, 2013) in drawing up plans to use the military to counteract the perceived threat of Trade Union action.

In many of its applications then 'keeping the lights on' becomes pretty much synonymous with the notion of energy security (Bradshaw, 2010; Winzer, 2012). While making most obvious sense when applied to electricity systems (as nearly all lighting is electrically powered), sometimes this specificity is lost and it is used as a metaphor of energy system security more generally.

What is the history of the idea of 'keeping the lights on'?

Keeping the lights on doesn't have a clear or obvious history, a 'first use' or similar. We might assume though that its appearance was associated with the establishment of electricity as a widespread and increasingly national infrastructure in more wealthy countries in the first half of the twentieth century (Hughes, 1993; Cubitt, 2013). Light was the first energy service that electricity was used to provide commercially, through lighting installations in high-end hotels and residences and stretches of electric street lamps in places such as Godalming in the UK, and Wabash in the USA. As local grids spread and eventually became joined together to form regional and then national grids, and in particular where municipal or national governments had direct responsibility for establishing and running these electricity systems, 'keeping the lights on' emerged as a responsibility of government and became symbolic of its competence.

It is significant that the breakdown of electricity supply has also come to be represented through a light-based idiom – 'blackout' has a variety of historical uses, including the wartime dousing of visible lights during air raids (Nye, 2010), but has come to be used to capture the most visible and obvious consequence of a power outage. In political terms blackouts are resolutely bad things to be avoided, the opposite, therefore, of the good thing to do – keeping the lights on. Power outages evidently do not only turn lights off. Much else that the modern world has come to depend on, such as information technology infrastructures, electronic payment systems, petrol pumps, tram, metro and electrified railway networks, can also fail to function (Shove, 2016; Walker, 2016). Light has become established and sustained as the commonplace idiom for talking about maintaining a secure electricity supply, despite the many ways in which it is now not only the light that matters.

What is wrong with 'keeping the lights on'?

So what is the problem? Surely keeping the lights on *is* a good thing to do? Taken literally artificial light is indeed an important energy service that brings well-being to people's lives in various ways (Petrova, 2017), enhancing their fundamental 'capabilities' (to use a language often applied to development issues) (Day, Walker and Simcock, 2016). Bringing electric light to rural communities in Africa, for example, is recognised as enabling people to work and study later into the evening, meaning they can enhance their education and livelihoods (Jacobson, 2007; Diouf, Pode and Osei, 2013). Streetlights and traffic lights make driving and streets safer, people feel more secure in well-lit urban spaces, and more prepared to walk and exercise outside, improving their health and well-being (Cerin et al., 2017). Artificial light matters undoubtedly in fundamental ways to making life better, so keeping light predictable and secure in its availability makes a lot of sense. If we extend beyond light to other energy services – such as heating, hot water and refrigeration – we can also spell out why they matter, and how they are also evidently fundamental to the basics of a normal, ordered, healthy and satisfactory everyday life (Walker, Simcock and Day, 2016). Surely we *do* need to avoid the many forms of interruption, disorder and sometimes chaos that ensues from power outages when the 'lights go out'?[2]

The problem is not with the objective, but rather how 'keeping the lights on' has come to be understood and how it is assumed to be realised – or how it *does work* in practice as a commonplace idiom or metaphor. The strongly embedded logic of the 'keeping the lights on' discourse is that if the problem is one of potential failure in supply – the electricity system in particular failing to be reliability secure in its operation – then that's where to focus attention, on supply. Core assumptions then follow that can be found reproduced across many examples. 'Keeping the lights on' means (i) replacing old power stations with new ones when they come to the end of their life, (ii) investing in additional capacity because energy demand is going to be higher in the future than it is now, (iii) securing new supplies of primary energy resource where these are needed to replace others that are running out or are too high in their carbon emissions, and (iv) investing in ways of generating power that are secure, reliable and predictable because that is what meeting demand needs. These are all actions that focus on supply, resources and generation, with demand the fixed variable that can only be followed not managed or reworked in any significant way. The extract below from an article titled 'Shale and nuclear are the way to keep the lights on' in the business pages of *The Telegraph* displays this supply-dominated logic:

> If we assume that coal will be phased out in the next few years then the burden on gas and nuclear only increases. We are totally dependent today as a country on gas and nuclear. There are no viable alternatives on any sensible time horizon. But, and it's a big but, we are fast running

out of gas in the North Sea and our nuclear fleet is ageing. North Sea gas production peaked in the year 2000, and is now running at less than 50% of its peak; in 10 years' time it will be at less than 20%. So we must choose between Russian imports and expensive LNG imports, or developing a shale industry of our own.

(Ratcliffe, 2016)

As in this article, which also argues that 'industry will wither and die' unless energy costs stay secure and competitive, the urgent focus on supply choices is often informed by a sense of foreboding and crisis – keeping the lights on will not be possible by other means, the route being advocated is necessary and vital, to do otherwise is to risk failure and political ignominy. Hence the seasonal pattern to media reporting in the UK; as winter approaches so does the potential for crisis if margins are cut too tight. And given the routine assumption that energy demand is always and inevitably going up in the future, there is crisis just over the horizon if the right decisions on supply are not made right now. Using a crisis-discourse has been long recognised as a political tactic in attempting to force or justify government action of a particular character and form (Jhagroe and Frantzeskaki, 2016), including in relation to 'peak oil' discourses (Bettini and Karaliotas, 2013; Huber, 2011). Arguably 'keeping the lights on' when deployed as part of an impending crisis narrative with the situation presented in those terms, can be read as an instance of intentional crisis-making that supports incumbent supply-side interests in the energy system.

How can the mentality of 'keeping the lights on' be challenged?

So what happens if those interests and the assumptions they are built on are challenged, most fundamentally by reorienting how the relationship between supply and demand is approached and how demand itself is thought about? Classic 'keeping the lights on' thinking sees demand as an aggregate block, to be met in its totality, essentially undifferentiated with all of the kilowatts of particular end uses adding up to make megawatts and gigawatts of total demand at any point in time (Patterson, 2007). There are daily and seasonal patterns in how these end uses add up, making load curves that supply needs to meet and constituting patterns that are to some degree predictable but *not* essentially changeable. The logic is thus one in which peak demand, which crisis narratives often focus on as the crunch point in time where demand exceeds supply capacity, will be what it will be; and supply has to be in place to meet it.

An alternative approach is to see demand not as a monolithic aggregate block, but made of an enormously differentiated set of uses of energy for different purposes as a part of many different everyday practices (Shove and Walker, 2014; Walker, 2014), and demand as continually being produced moment by moment – a living and breathing entity in effect, complex and interwoven with how society functions on an ongoing basis. Focusing on peak demand in these terms can lead us to ask what are peaks in fact made up of (Torriti et al., 2015), why a peak is produced when it is, and crucially whether some of what makes it up could happen at a different point in time. If armed with this knowledge, the dynamics of demand can be intervened in to spread or 'shave' the peak, by making particular uses of energy more flexible and responsive in their timing to the needs of the energy system (Torriti, 2016), then we have an alternative way of avoiding the 'crunch point' – managing demand to fit with supply capacity rather than supply always and forever following and meeting demand.

This alternative approach to keeping the lights on is increasingly becoming recognised as having an important place in the management of electricity grids (Torriti and Grunewald, 2014). For example, a suite of demand response schemes are being run by National Grid in the UK[3] with the aim of incentivising industrial and commercial energy users to essentially turn energy-using devices off at peak times in order to help keep the wider electricity system in balance (which includes avoiding high cost and high carbon generation); and there is much attention also being given to finding flexibility in the temporal patterns of domestic energy demand once smart metering technologies are in place. The implication is though that turning some things off is still about keeping a vast multiplicity of other things switched on.

Shifting demand around in time is therefore only a starting point in challenging classic 'keeping the lights on' thinking. The wider insight is that by approaching demand differently, disaggregating and not taking it for granted, opportunities to intervene are revealed. This applies to longer term trajectories in which the 'keeping the lights on' crisis is presented as a problem of how to keep up with ever rising electricity demand. Rather than just taking the assumption of ever rising demand for granted – indeed as a good thing symbolic of strength and economic power – approaching demand differently opens up questions about how its trajectories are both internally differentiated and potential sites of intervention. Questions can then be asked about the value and need for particular uses of energy and the normalised ways of living they are sustaining.

Even though, as argued earlier, a range of energy services can be seen as enabling the basics of well-being, whether and to what degree energy use, in all of its contemporary variety, really needs to be integral to living well is an open question. Evidently many of the ways in which we have become apparently energy-dependent are about going far beyond the basics of well-being. Considering light as just one example, yes modern lighting technologies provide for the basics of illumination in people's homes, in workplaces, on streets and so on. But it has also become an integral part of interior design and the selling of products in shops, both of which have proliferated the number of light fittings and illuminated bulbs. Many lights are now kept on in the daytime even when natural light is plentiful, more buildings are floodlit than in the past, sports grounds are ablaze into the evening. The use and purpose of artificial light has evidently been chang-ing over time and we are undoubtedly keeping exponentially many more lights on than we used to, counteracting to some degree the significant gains in bulb efficiency (Boardman, 2014; Jensen, 2017) that have been achieved. So does 'keeping the lights on' just mean continually meeting all of this evolving demand for electricity, all of the time, regardless? And how about all of the other examples of proliferating and escalating pat-terns of energy demand that we could identify, information technologies being a prime example (Hazas, 2015)? Surely questions can and should be asked about the need for energy and the scope for using much less, rather than just accepting the dynamics of demand as they materialise now and into the future?

Keeping energy systems working and avoiding failure is important, but we need to look at how to achieve this beyond building more supply to meet an unquestioned demand. Starting with asking what in fact energy systems are working for would be a good first step, opening up a question that is systematically ignored by classic 'keeping the lights on' thinking.

Notes

1 See Millán et al. (2003) on Latin America; Nahzi (2016) on Kosovo; EAC (2009) on the US; Boren (2016) on the UK; Anderson (2015) on the Middle East; and Shell (undated) on the Philippines.
2 In many less economically developed countries blackouts can be a relatively regular and 'normal' occurrence, meaning the degree of immediate disruption and chaos is lessened, although the social and economic consequences of being without elec-tricity can still be severe.
3 See the National Grid's 'Power Responsive' initiative (http://powerresponsive.com).

Further reading

Torriti, J. (2016). *Peak energy demand and demand side response*. London: Routledge.
Walker, G. (2014). The dynamics of energy demand: Change, rhythm and synchronicity. *Energy Research & Social Science*, 1, pp. 49–55.

Walker, G. (2016). De-energising and de-carbonising society: Making energy (only) do work where it is really needed. Retrieved from www.foe.co.uk/sites/default/files/downloads/de-energising-society-102383.pdf [accessed 15 February 2019].

References

Anderson, B. (2015). Keeping the lights on in Lebanon, Iraq, Jordan. Retrieved from www.bbc.com/capital/story/20150616-blackouts-bribes-and-red-tape [accessed 15 February 2019].

Bettini, G. and Karaliotas, I. (2013). Exploring the limits of peak oil: Naturalising the political, de-politicising energy. *The Geographical Journal*, 179, pp. 331–341.

Boardman, B. (2014). Low-energy lights will keep the lights on. *Carbon Management*, 5, pp. 361–371.

Boren, Z. (2016). Yes, the UK will keep the lights on this winter. *Unearthed*, 14 October. Retrieved from https://unearthed.greenpeace.org/2016/10/14/uk-keeping-lights-done-yet [accessed 15 February 2019].

Bradshaw, M. J. (2010). Global energy dilemmas: A geographical perspective. *Geographical Journal*, 176, pp. 275–290.

Cerin, E., Nathan, A., Van Cauwenberg, J. Barnett, D.W., Barnett, A. and Council for Environment and Physical Activity (2017). The neighbourhood physical environment and active travel in older adults: A systematic review and meta-analysis. *International Journal of Behavioral Nutrition and Physical Activity*, 14(15).

Cubitt, S. (2013) Electric light and electricity. *Theory, Culture and Society*, 30, pp. 309–323.

Day, R., Walker, G. and Simcock, N. (2016). Conceptualising energy use and energy poverty using a capabilities framework. *Energy Policy*, 93, pp. 255–264.

Diouf, B., Pode, R. and Osei, R. (2013). Initiative for 100% rural electrification in developing countries: Case study of Senegal. *Energy Policy*, 59, pp. 926–930.

EAC (2009). *Keeping the lights on in a new world*. Washington, DC: Electricity Advisory Committee. Retrieved from https://energy.gov/sites/prod/files/oeprod/DocumentsandMedia/adequacy_report_01-09-09.pdf [accessed 15 February 2019].

Guardian (2016). The Guardian view on energy policy: keeping the lights on is what governments are for. *The Guardian*, 20 March. Retrieved from www.theguardian.com/commentisfree/2016/mar/07/the-guardian-view-on-energy-policy-keeping-the-lights-on-is-what-governments-are-for [accessed 23 November 2018].

Hazas, M. (2015). Society pushes to go faster, but data binges carry environmental costs. Retrieved from https://theconversation.com/society-pushes-to-go-faster-but-data-binges-carry-environmental-costs-36672 [accessed 23 November 2018].

Huber, M. T. (2011). Enforcing scarcity: Oil, violence, and the making of the market. *Annals of the Association of American Geographers*, 101, pp. 816–826.

Hughes, T. (1993). *Networks of power: Electrification in Western society, 1880–1930*. Baltimore, MD: Johns Hopkins University Press.

Jacobson, A. (2007). Connective power: Solar electrification and social change in Kenya. *World Development*, 35, pp. 144–162.

Jensen, C. (2017). Understanding energy efficient lighting as an outcome of dynamics of social practices. *Journal of Cleaner Production*, 165, pp. 1097–1106.

Jhagroe, S. and Frantzeskaki, N. (2016). Framing a crisis: Exceptional democracy in Dutch infrastructure governance. *Critical Policy Studies*, 10, pp. 348–364.

Marsden, S. (2013). Thatcher made secret plans to bring in the military during the miners' strike. *Daily Telegraph*, 1 August. Retrieved from www.telegraph.co.uk/news/politics/margaret-thatcher/10213447/Thatcher-made-secret-plans-to-bring-in-the-military-during-the-miners-strike.html [accessed 23 November 2018].

Millán, J., von der Fehr, N.-H. M., Ayala, U., Walker, I., Benavides, J., Fundación S., Rufin, C. and Brown, A. (2003). *Keeping the lights on: Power sector reform in Latin America*. Washington, DC: IDB. Retrieved from https://publications.iadb.org/handle/11319/311 [accessed 15 February 2019].

Nahzi, F. (2016). Keeping the lights on in Kosovo. 21 July. Retrieved from www.huffingtonpost.com/fron-nahzi/keeping-the-lights-on-in-_b_11057686.html [accessed 15 February 2019].

Nye, D. E. (2010). *When the lights went out: A history of blackouts in America*. Cambridge MA: MIT Press.

Patterson, W. (2007). *Keeping the lights on: Towards sustainable electricity*. London: Chatham House and Earthscan.

Petrova, S. (2017). Illuminating austerity: Lighting poverty as an agent and signifier of the Greek crisis. *European Urban and Regional Studies*, 25(4), pp. 360–372.

Ratcliffe, J. (2016). Shale and nuclear are the way to keep the lights on. *The Telegraph*, 14 October. Retrieved from www.telegraph.co.uk/business/2016/10/14/shale-and-nuclear-are-the-way-to-keep-the-lights-on [accessed 14 October 2016]

Shell (undated). Keeping the lights on in the Philippines. Retrieved from www.shell.com/inside-energy/keeping-the-lights-on-in-the-philippines.html [accessed 23 November 2018].

Shove, E. (2016). When is the power really on? Retrieved from www.demand.ac.uk/20/01/2016/reflections-on-the-lancaster-power-cuts-of-december-2015 [accessed 23 November 2018].

Shove, E. and Walker, G. (2014). What is energy for? Social practice and energy demand. *Theory, Culture & Society*, 31, pp. 41–58.

Tennessee Aquarium (2015). Meet the Tennessee Aquarium's tweeting electric eel. Retrieved from www.tnaqua.org/newsroom/entry/meet-the-tennessee-aquariums-tweeting-electric-eel [accessed 20 November 2018]

Torriti, J. (2016). *Peak Energy Demand and Demand Side Response*. London: Routledge.

Torriti, J., Druckman, A., Anderson, B., Yeboah, G. and Hanna, R. (2015). Peak residential electricity demand and social practices: Deriving flexibility and greenhouse gas intensities from time use and locational data. *Indoor and Built Environment*, 24, pp. 891–912.

Torriti, J. and Grunewald, P. (2014). Demand side response: Patterns in Europe and future policy perspectives under capacity mechanisms. *Economics of Energy & Environment Policy*, 3, pp. 87–105.

Walker, G. (2014). The dynamics of energy demand: Change, rhythm and synchronicity. *Energy Research & Social Science*, 1, pp. 49–55.

Walker, G. (2016). Lancaster with and without power: Reflections of the dynamics of energy, mobility and demand. Retrieved from www.demand.ac.uk/20/01/2016/reflections-on-the-lancaster-power-cuts-of-december-2015 [accessed 23 November 2018].

Walker, G., Simcock, N. and Day, R. (2016). Necessary energy uses and a minimum standard of living in the United Kingdom: Energy justice or escalating expectations?. *Energy Research & Social Science*, 18, pp. 129–138.

Winzer, C. (2012). Conceptualizing energy security. *Energy Policy*, 46, pp. 36–48.

9

PROMOTING SMART HOMES

Mike Hazas and Yolande Strengers

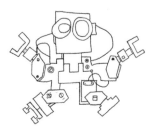

A smart robot is an artificial intelligence system that can learn from its environment and build on its capabilities based on that knowledge. Smart robots can collaborate with humans, working along-side them and learning from their behaviour. Some people think smart homes will be able to do the same.

Introduction

Once upon a time, in the not-too-distant future, a family lived happily in their smart home, equipped with seamless and easy automation, and digital control. They saved money and energy through on-site generation which powered their smart appliances when available. They automated their lights, climate control and other devices to ensure energy wasn't wasted. They had more time for themselves and their friends, and better entertainment possibilities than ever before.

Or so the story goes.

Proponents of so-called 'smart home technologies' say that embedding digital sensing, automation and control into everyday life will reduce energy consumption with little or no effort. At the same time, they argue, our lives will be more comfortable, convenient and pleasurable. Smart home technologies – energy meters, thermostats, lighting and security systems, and digital voice assistants such as Google Home and Amazon Alexa – are the celebrated heroes and heroines in a story in which technology will enable households to save energy in a low carbon society. 'Smartness' is also widely inscribed into the design and marketing of a whole host of household consumables: laptops, smartphones, TVs and hi-fi systems, kettles, refrigerators, electric vehicles and microgeneration (solar panels and wind turbines).

Even though visions for 'smart' are variable and fluid (Darby, 2018) the term generally refers to the intelligence embedded in specific technologies, many of which are Internet-enabled. This allows for automation or 'smart control' of devices, usually through a tablet or smart phone app, but increasingly through voice control. Smart homes consequently depend on information and communication networks to connect appliances and systems to one another and enable remote access or control of a variety of home services (Aldrich, 2003; Darby, 2018). However, as Darby (2018, p. 142) notes there 'has often been a sometimes confusing mix' of two dominant smart home ambitions: the first being focused on 'luxury home living with a modern flavour and a tang for efficiency'; and the second oriented towards balancing supply and demand on the electricity network, through new forms of efficiency and automation.

These competing narratives (luxury and energy reduction) are potentially important in that analysts predict that the smart home market will increase 15 per cent by 2022 (Zion, 2017), growing to approximately US$53 billion, even though uptake of smart home technologies is widely acknowledged as being much slower than expected (see Katuk et al., 2018). In the meantime, the energy-related impact of current 'smart' home devices is ambivalent.

This is an emerging field, but overall, the deployment and use of smart home devices has not been strongly linked to reductions in energy consumption. Some specific applications of energy-related smart home devices have led to reductions in energy demand, or have helped modify the timing of demand (e.g. through energy feedback (Buchanan, Russo and Anderson, 2014) or home automation). However, other applications have been linked to increased direct energy requirements (Wilson and Hargreaves, 2017; Tirado Herrero, Nicholls and Strengers, 2018) and additional indirect energy demand involved in sending, storing and processing large amounts of data in external servers (Morley, Widdicks and Hazas, 2018, pp. 128, 130).

Promoting the smart home

Given these uncertainties, it is important to learn more about where the idea that smart homes save energy comes from. We identify three main proponents. The first includes governments, energy policy makers, and energy industry stakeholders who have increasingly embraced the idea that smart technologies will help households save energy and address a range of pressing issues including climate change, energy security, peak electricity demand and energy poverty (Strengers, 2013; Tirado Herrero, Nicholls and Strengers, 2018). This is a part of a broader twenty-first-century agenda of 'smartification' that has taken hold across various portfolios of public policy, and is emblematic of the so-called Fourth Industrial Revolution (WEF, 2017). The popularity of smart home technologies as a response to energy affordability, security and efficiency closely reflects a broader trend towards 'technology solutionism' that pervades neoliberal governments (Morozov, 2013).

Smart meters, grids, homes and cities are touted as 'revolutionary' (European Commission, 2006), 'transformative' (Wimberly, 2011) and 'disruptive' (WEF, 2017) in policy documents, referring to their assumed ability to significantly change and improve the future. In the residential energy sector, this transformation is expected to follow from automating consumption, generating cleaner and more efficient energy on-site, and/or providing residents with up-to-date energy data so they can make 'informed choices'.

The second group includes smart home technology companies, which often promise energy savings from using their devices (Strengers et al., 2016). However, the benefit of energy savings plays less prominent a role in both product information and marketing. More central are the goals of enhancing convenience and creating a cosy atmosphere and luxurious environment – and in current realisations, these tend to consume more energy, rather than less.

Closely related are a third group of smart home supporters or advocates, including industry analysts, technology developers, and academic researchers, who are writing and talking about the energy-saving benefits of smart home technologies in highly positive ways (as critiqued by Strengers and Nicholls, 2017; Wilson, Hargreaves and Hauxwell-Baldwin, 2014). Most commonly these positive stories are written into media articles and marketing materials. They tend to overemphasise the benefits of smart home technologies, and overlook the complexities and problems of achieving the energy-saving vision.

We identify three key features of the common narrative that smart technologies are critical to achieving energy-savings in the home, and then discuss two examples that illustrate some of the problems with this idea.

Smart technologies make energy saving easy

An overarching assumption is that smart home technologies make it easy to save energy. Energy savings are assumed to be simple to achieve by using devices that turn things off when not in use, relying on sensors or monitors to control certain daily functions (e.g. raising or lowering blinds and opening or closing window), or automating and 'learning' inhabitants' daily routines to customise devices around their needs.

Smart technologies are easy to use

Beyond making energy saving easy, smart devices are marketed more generally as being easy to use, 'straight out of the box'. Off-the-shelf, 'plug-and-play' or 'set-and-forget' devices promote the idea that it's as simple as plugging in a device and seamlessly 'playing' with it – or forgetting about it, as noted by Harper-Slaboszewicz, McGregor and Sunderhauf (2012). Smart technologies are promoted as being 'intuitive' to install, configure and enjoy. In practice, recent 'usability' studies have identified issues with technology compatibility (e.g. with smart phone models), difficulties downloading and using apps, access to (and affordability of) a reliable wifi connection and unfamiliarity with how smart phones and apps work (particularly for older households) (Nicholls, Strengers and Tirado, 2017).

Smart technologies make it possible to do more with less

A third feature of the smart home vision, particularly as advocated by smart home companies, is that energy-savings are an added side-benefit of an enhanced and more luxurious lifestyle. The idea promoted by smart home companies and advocates is that inhabitants can enjoy a higher quality of life (more) while using less energy (less). In other words, they can have more (comfort, convenience, leisure time, entertainment, security etc.) and simultaneously reduce their energy consumption.

How do smart home technologies work out in practice?

In this section we introduce two present-day examples of smart home technology relating to the practices of heating and cooling; and lighting. We show how these reproduce all three aspects of the smart home story introduced above, and we show that in each case the upshot is likely to be an increase and not a decrease in energy demand.

Smart heating and cooling

Indoor heating and cooling account for significant proportions of energy demand and there is a long history of effort to reduce consumption through efficiency measures such as insulation and draughtproofing, and more recently programmable thermostats. Smart thermostats are designed as a direct replacement (*easy to use*) for the conventional or programmable thermostat; these are normally situated on the wall, in a hallway or living room. The most prominent example is the Learning Thermostat made by Nest Labs (a subsidiary of Google); the device is often called 'the Nest'. At minimum, smart thermostats incorporate a digital thermometer and an infrared motion sensor which is used to infer occupancy. The idea is that they automatically limit heating and cooling to times when the house is occupied (*easy to save*). Most smart thermostats can be controlled from outside the home, using a web browser or smartphone app. This is often marketed as providing enhanced convenience and peace of mind, while simultaneously saving energy by avoiding potential wastage (*more with less*).

However, even with per-room occupancy and temperature information and tailored algorithms, 'smartly' managed heating and cooling can result in an increase energy consumption depending very much on the house, and how thermostats were previously set. As Scott et al. (2011) found, in some instances energy savings were wiped out by an increased level of service provision: for instance, automatic heating of the home every time the system anticipated it would be occupied (see Chapters 3 and 4). Others have concluded that the algorithms are over-responsive, taking occasional motion detection and feeding this into the automatically determined schedule (Yang, Newman and Forlizzi, 2014).

While Nest sometimes claim large savings from using their devices, the comparison is often with always-on heating or cooling – following the industry logic that programmable timer thermostats are not used correctly anyway (Nest Labs, 2015, pp. 2–3). When compared only to programmable thermostats, the heating savings are about 7 per cent, with negligible savings for cooling (ibid., p. 11). However, in many parts of the world, such as parts of Australia, Europe and Southeast Asia, fully thermostatically controlled homes are still uncommon, and many people experience more 'natural' conditions or 'adaptive' modes of comfort in their home (e.g. Chappells and Shove, 2005; Strengers and Maller, 2011; Winter, 2013).

If these households were to adopt the Nest as an 'efficient' heating or cooling system, they would also be adopting an energy-intensive ideal of comfort which aims to produce thermostatically stable environments. More broadly, the promotion of automated devices like the Nest reproduces ideas about thermostatically controlled home environments at the expense of other possibilities including local or culturally specific variations such as opening windows, using fans, wearing warming/cooling clothing, taking a cold/hot shower, or spending time in public cool spaces and places.

Smart lighting

Compared to traditional incandescent lighting, compact fluourescent and most recently LED lights have led to lower energy intensity at each light fixture or socket – by simply replacing the bulb. However, evidence is strong that this efficiency gain is at least partly negated by the increase in the number of lights installed in buildings worldwide (Shellenberger and Nordhaus, 2014). The various functionalities of smart LED lighting also have the potential to both decrease *and* increase consumption.

The most common smart light bulbs and lighting appliances are simply plugged in and then controlled through a smart phone app, tablet, and/or wall switch (*easy to use*). These apps can be used to turn the lights on or off, group lights together to create 'scenes' (see below), set the dimming level, and in many cases set the colour scheme of the light or light group. Smart lighting also provides opportunities to save energy by turning off groups of lights in one touch (or voice command), checking to see if any lights have been accidentially left on in the home (and turning them off), and automatically scheduling lights to come on and off. Some more advanced systems provide 'goodnight' and 'away' scenes which turn off all lights and unnecessary appliances in the house when going to bed or leaving the home, thereby avoiding the potential for wasted or unnecessary energy demand.

However, smart lights are also marketed as ways of providing new sensory experiences in the home (*more with less*). 'Mood lighting', available through lighting scenes, is meant to induce or enhance inhabitants' feelings, for example of invigoration or calm. Occupants are expected to use lighting to enhance the mood for meditation, housework or a party through the creation of scenes. As part of these possibilities, lighting may be incorporated into walls, under beds, or behind cupboards and benches via LED strip-lighting and back-lighting where previously lighting didn't exist or wasn't deemed necessary. Smart lighting is thus closely tied to generating 'a luxurious aesthetic experience' (Strengers and Nicholls, 2017, p. 2).

Additionally, smart lights require a wireless network connection; often this is through wifi, but some smart lights require a dedicated networked 'gateway' that must be plugged in elsewhere in the home. The smart light listens for instructions on the network *all the time*, which can easily make up half of the total energy consumption of the light, with the other half used to supply the actual light source (International Energy Agency, 2016).

Challenging assumptions about smart home technologies

The energy-saving smart home masks possible rises in energy intensity, due to changes in fidelity, convenience, comfort or the inclusion of enhanced sensory experiences. While savings are possible, it is unclear whether smart technologies will actually reduce household energy consumption, and overall demand. As we explain below, assumptions on which 'smart' narratives depend may have surprising and counterproductive consequences.

First, the rhetoric on smart home technologies assumes that sensors and algorithms are capable of capturing and reasoning about the salient features of how inhabitants live in homes. For example, it is assumed that learning thermostats can successfully detect when no one is in the house. However, this doesn't always happen. For example, a brief return home can result in a short interval of heating or cooling being erroneously added to the regular schedule (Yang, Newman and Forlizzi, 2014). And it can take up to 'a couple of hours' after departure before the Nest thermostat will automatically lower heating or cooling (Nest Labs, undated).

Second, familiar narratives about smart home technologies assume that negligible amounts of energy are required to run smart devices, connect them to digital networks, and store and process their data. In most examples this is simply a case of failing to acknowledge these additional energy demands. When energy savings figures are claimed, they do not typically take into account the additional household electricity demand that is generated by the smart home technology (e.g. the networked standby power of smart LED light bulbs), nor the energy needed for transmission, processing and storage of data beyond the home (e.g. the Internet and data centres – see Morley et al., 2018).

Finally and related to the second assumption, product literature and smart home proponents promote the continual improvement of smart home technologies through software updates. This presupposes a constant, reliable Internet connection being available in the home. For the Nest learning thermostat, this connection must be over wifi which tends to be more difficult to manage and troubleshoot (Nicholls, Strengers and Tirado, 2017). Perhaps just as crucial for energy demand, the sum total of software updates currently contributes to significant load on Internet communications networks: one study showed that software updates accounted for much as 10 per cent of network traffic for phones and tablets (Widdicks et al. 2017; see also de Decker, 2017). By leaving this out of the discussion, those who advocate smart home technologies overlook what is likely to be a major consideration given the billions of smart home devices that are expected to be deployed in the next twenty years.

Since smart technologies are (in theory) flexible, other scenarios are possible. There is consequently scope for fundamentally reimagining and revisioning their role in enabling alternative, less energy demanding ways of life. For instance, rather than silently maintaining set temperatures at occupied times, smart thermostats could focus more on ventilation and shading than on mechanical heating and cooling. Similarly, lighting systems could seek to make the most of ambient light from windows (perhaps using motorised blinds), or from traditional lights which are already on. Rather than relying on always-on Internet connections and remote data centres, all smart devices could be designed to stay offline and autonomous most of the time to conserve energy both inside and beyond the home. All are possible but as is now obvious, moving in these directions challenges some of the core assumptions on which present discourses depend.

In conclusion, narratives about smart homes mask and hide the multiple forms of energy consumption that such technologies enable, while simultaneously legitimisating and naturalising always-on and constantly updating devices and software. While there are benefits (in efficiency, automation and demand management) we need to consider the 'full story' in order to appreciate the opportunities and problems involved in turning to the smart home as a means of lowering energy demand.

Further reading

Darby, S. J. (2018). Smart technology in the home: time for more clarity. *Building Research & Information*, 46(1), pp. 140–147.

Tirado Herrero, S., Nicholls, L. and Strengers, Y. (2018). Smart home technologies in everyday life: Do they address key energy challenges in households? *Current Opinion in Environmental Sustainability*, 31, pp. 65–70.

Wilson, C., Hargreaves, T. and Hauxwell-Baldwin, R. (2017). Benefits and risks of smart home technologies. *Energy Policy*, 103, pp. 72–83.

References

Aldrich, F. K. (2003). Smart homes: past, present and future. In R. Harper (ed.), *Inside the Smart Home*. London: Springer-Verlag, pp. 17–39.

Buchanan, K., Russo, R. and Anderson, B. (2014). Feeding back about eco-feedback: How do consumers use and respond to energy monitors? *Energy Policy*, 73, pp. 138–146.

Chappells, H. and Shove, E. (2005). Debating the future of comfort: Environmental sustainability, energy consumption and the indoor environment. *Building Research and Information*, 33(1), pp. 32–40.

Darby, S. J. (2018). Smart technology in the home: time for more clarity. *Building Research & Information*, 46(1), pp. 140–147.

De Decker, K. (2017). Rebooting energy demand: Automatic software updates. Retrieved from www.demand.ac.uk/rebooting-energy-demand-automatic-software-updates [accessed 6 Sep 2018].

European Commission (2006). *European SmartGrids technology platform: Vision and strategy for Europe's electricity networks of the future*. Luxembourg: Office for Official Publications of the European Communities.

Harper-Slaboszewicz, P., McGregor, T. and Sunderhauf, S. (2012). Customer view of smart grid – set and forget? In F. P. Sioshansi (ed.), *Smart Grid*. Boston: Academic Press, pp. 371–95.

International Energy Agency (2016). *Solid state lighting annex: Task 7: Smart lighting – new features impacting energy consumption.* Paris: International Energy Agency. Retrieved from https://ssl.iea-4e.org/files/otherfiles/0000/0085/SSL_Annex_Task_7_-_First_Report_-_6_Sept_2016.pdf [accessed 23 November 2018].

Katuk, N., Ku-Mahamud, K. R., Zakaria, N. H. and Maarof, M. A. (2018). Implementation and recent progress in cloud-based smart home automation systems. Paper presented at 2018 IEEE Symposium on Computer Applications & Industrial Electronics (ISCAIE), 28–29 April 2018, Penang, Malaysia. Retrieved from http://repo.uum.edu.my/24423 [accessed 15 February 2019].

Morley, J., Widdicks, K. and Hazas, M. (2018). Digitalisation, energy and data demand: The impact of Internet traffic on overall and peak electricity consumption. *Energy Research & Social Science*, 38, pp. 128–137.

Morozov, E. (2013). *To save everything click here: Technology, solutionism and the urge to fix problems that don't exist.* London: Penguin Books.

Nest Labs (2015). Energy savings from the Nest Learning Thermostat: Energy bill analysis results. Retrieved from http://downloads.nest.com/press/documents/energy-savings-white-paper.pdf [accessed 4 Sep 2018].

Nest Labs (undated). Troubleshooting Eco Temperature switching when your Nest Thermostat is installed on the Nest Stand. Retrieved from https://nest.com/uk/support/article/Troubleshooting-Eco-Temperature-switching-when-your-Nest-Thermostat-is-installed-on-the-Nest-Stand [accessed 23 November 2018].

Nicholls, L., Strengers, Y. and Tirado, S. (2017). *Smart home control: Exploring the potential for off-the-shelf enabling technologies in energy vulnerable and other households.* Melbourne: Beyond Behaviour Research Programme, Centre for Urban Research (CUR), RMIT University. See table 10 in particular. Retrieved from https://researchbank.rmit.edu.au/view/rmit:44455 [accessed 30 August 2018].

Scott, J., Brush, A. J., Krumm, J., Meyers, B., Hazas, M., Hodges, S. and Villar, N. (2011). PreHeat: Controlling home heating using occupancy prediction. In *Proceedings of the 13th International Conference on Ubiquitous Computing*, pp. 281–290. New York: ACM. doi: 10.1145/2030112.2030151

Shellenberger, M. and Nordhaus, T. (2014). The problem with energy efficiency. *The New York Times*, 8 October. Retrieved from https://nyti.ms/1EwiHQ7 [accessed 23 November 2018].

Strengers, Y. (2013). *Smart energy technologies in everyday life: Smart Utopia?* London: Palgrave Macmillan.

Strengers, Y. and Maller, C. (2011). Integrating health, housing and energy policies: The social practices of cooling. *Building Research & Information*, 39(2), pp. 154–168.

Strengers, Y. and Nicholls, L. (2017). Convenience and energy consumption in the smart home of the future: Industry visions from Australia and beyond. *Energy Research & Social Science*, 32, pp. 86–93.

Strengers, Y., Nicholls, L., Owen, T. and Tirado, S. (2016). *Smart home control devices: Summary and assessment of energy and lifestyle marketing claims.* Melbourne: Centre for Urban Research (CUR), RMIT University. Retrieved from http://apo.org.au/node/72305/cite [accessed 30 August 2018].

Tirado Herrero, S., Nicholls, L. and Strengers, Y. (2018). Smart home technologies in everyday life: Do they address key energy challenges in households? *Current Opinion in Environmental Sustainability*, 31, pp. 65–70.

WEF (2017). *The future of electricity: New technologies transforming the grid edge.* Geneva: World Economic Forum.

Widdicks, K., Bates, O., Hazas, M., Friday, A. and Beresford, A. R. (2017). Demand around the clock: Time use and data demand of mobile devices in everyday life. In *Proceedings of the 2017 CHI Conference on Human Factors in Computing Systems*, pp. 5361–5372. New York: ACM.

Wilson, C., Hargreaves, T. and Hauxwell-Baldwin, R. (2014). Smart homes and their users: A systematic analysis. *Personal and Ubiquitious Computing*, 19(2), pp. 463–476.

Wimberly, J. (2011). *EcoPinion consumer cents for smart grid survey report*, issue 12. Washington: EcoAlign. Retrieved from www.pointview.com/data/files/3/2656/2017.pdf [accessed 30 August 2018].

Winter, T. (2013). An uncomfortable truth: Air-conditioning and sustainability in Asia. *Environment and Planning A*, 45(3), pp. 517–531.

Yang, R., Newman, M. W. and Forlizzi, J. (2014). Making sustainability sustainable: Challenges in the design of eco-interaction technologies. In *Proceedings of the SIGCHI Conference on Human Factors in Computing Systems*, pp. 823–832. New York: ACM. doi: 10.1145/2556288.2557380

Zion (2017). *Smart home market (smart kitchen, security & access control, lighting control, home healthcare, hvac control and others): Global industry perspective, comprehensive analysis, and forecast, 2016–2022.* Maharashtra: Zion Market Research.

PART IV
Policies

10

THE ENERGY TRILEMMA

Jenny Rinkinen and Elizabeth Shove

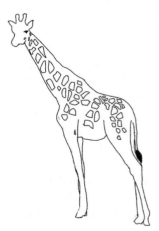

Like many other animals, real and imagined, giraffes have four legs, not three. The energy trilemma refers to tensions between three poles: energy security, affordability and decarbonisation. Is energy demand a missing dimension?

Introduction

The 'energy trilemma' is a term used to describe the policy challenge of simultaneously responding to the potentially competing goals of energy security, energy affordability and low carbon energy supply. The fact that the energy trilemma has become a powerful rhetorical device across policy and research is not surprising in that it is consistent with other prevalent policy discourses. For example, the three dimensions of the trilemma (security, affordability and low carbon supply) resonate with the suggestion that there are three pillars of sustainability – economic, social and environmental – an idea that has figured strongly in policy discussions since the Bruntdland report of 1987.

The trilemma also resonates with popular discourses of clean energy, green growth and efficiency (HM Government, 2017). For example, clean energy policies are not only designed to cut pollution, but to also boost employment and national competitiveness, and help ensure a reliable energy supply.

In addition, and because the trilemma is frequently cited in support of efforts to diversify primary energy supply and electricity generation (World Energy Council, 2016), the concept is strongly associated with a technological view of provision. For example, Blumsack and Fernandez (2012) argue that the smart grid is vital in addressing the trilemma.

Although it is said to provide a holistic view of energy systems, we contend that the trilemma is in fact lopsided and partial. In focusing on features of supply it overlooks fundamental questions about the scale and dynamics of demand. In effect, talk of the energy trilemma, and strategies and policies informed by it, take energy demand and the practices on which demand depends for granted.

In using the energy trilemma to frame the terms of energy policy, and positioning it as a 'basis for the prosperity and competitiveness of individual countries' (World Energy Council, 2016, p. 2), institutions such as the World Energy Council direct attention away from questions about demand and how it evolves. For example, discussions about how the goals of security, affordability and decarbonisation might be balanced rarely refer to the extent and character of demand. Behind the scenes, there are numerous tacit assumptions about demand – after all, the meaning of 'affordability' and also 'security' are in the end related to the quantities of energy involved. But because these understandings are almost never explicit, debates about how the trilemma might be managed almost always focus on the supply side and on the means through which demands are met (Shove and Walker, 2014).

Amongst other things, this approach supposes that systems and technologies of supply are not, in themselves, implicated in the constitution of energy demand. It also excludes serious analysis of how demand changes and what this means for both the scale and the character of the challenges represented in the trilemma.

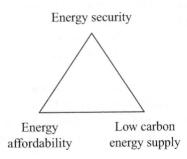

FIGURE 10.1 The energy trilemma.

A history of the energy trilemma

The *Merriam-Webster Dictionary* definition of a trilemma is that it is 'a state of things in which it is difficult to determine which one of three courses to pursue'.[1] The concept of a 'trilemma' is used in disciplines such as philosophy and business to refer to a choice between three options, each of which is (or appears) unacceptable or unfavourable in that it might compromise other goals. Often the trilemma is mentioned in discussions about how to find an acceptable balance between three options.

In the energy field, the electricity supplier E.ON was one of the first to describe the challenge of balancing the sometimes conflicting goals of energy security, energy affordability and low carbon energy supply as the 'energy trilemma' (E.ON UK, 2008, cited in Boston, 2013). Used in this way, it indicates the existence of multiple considerations, and the complexity of the energy system as a whole.

A number of countries explicitly refer to the trilemma in their energy policies and plans. For example, in the UK, the objectives of the Electricity Market Reform Bill are clearly spelled out with reference to the trilemma, and in the business sector, efforts to encourage investment in lower carbon products and technologies are justified in the same terms (HM Government, 2014). The trilemma has also been used in various policy analyses such as those by Sautter, Landis and Dworkin (2008) and Gunningham (2013), and by UK research councils (Sharick, 2015). However, the details of exactly how one node or pole relates to another are unclear: more or less effort or investment at one point of the triangle does not translate equally, or have any necessary effect on any of the other dimensions.

Despite these complexities, there have been some attempts to describe policy interventions with reference to a trilemma 'index'. Since 2011, the World Energy Council has been particularly active in promoting the concept, publishing annual reports on what it describes as the 'World Energy Trilemma'. The latest report, *Defining measures to accelerate the energy transition*, evaluates the performance of different countries using a 'trilemma index'. This index, discussed at greater length below, is presented as an important tool for assessing progress towards lower carbon energy transitions (World Energy Council, 2017).

To understand how the concept of the trilemma influences plans, ambitions and forms of evaluation we need to take a closer look at each of the three dimensions: security, affordability and lower carbon supply.

Security

First, security. Following the oil crises of the 1970s, soaring energy prices and other uncertainties prompted energy-importing countries to re-evaluate levels of self-sufficiency and the security of their energy supplies. The availability of different fuel reserves and their affordability are central to the International Energy Agency's definition of energy security 'as the uninterrupted availability of energy sources at an affordable price' (International Energy Agency, 2014). According to the World Energy Council, energy security is about the management of primary energy supply from domestic and external sources (at any scale), reliability of energy infrastructure, and the ability of energy providers to meet current and future demand (World Energy Council, 2016). But what does this really mean?

In practice, specifying and planning for 'security of supply' is much more complicated than these broad statements suggest. For example, decisions have to be made about exactly how much gas or oil should be kept in reserve, and there are judgements to be made about how long these reserves should last. Questions of resilience – often discussed in terms of the need to 'keep the lights on' (see Chapter 8) – also influence major investment strategies (for example, building the Hinkley Point nuclear plant in the UK). In general, a preoccupation with security of supply favours investment in the technologies and resources needed to meet 'baseload', often at the expense of developing other more flexible forms of infrastructure.

Meanwhile, and over the longer run, the social and political significance of security evolves. For instance, changes in the energy mix (such as those associated with the need to decarbonise forms of energy supply), change the profile of security concerns, and shifts in technology (such as the plummeting cost of solar energy; the decreasing competitiveness of combined cycle gas turbines) have the same effect. This is a circular process. Previous concerns about security favour some kinds of technological investment and not others: so shaping both the context and the definition of present concerns, and their salience in contemporary energy policy.

Affordability

Energy affordability is often seen as the social or the economic aspect of the trilemma. It refers to the availability of affordable energy across the population, and as such rests on some tacit understanding of how much energy people need. In broad terms, 'affordability' describes the extent to which members of a population can access electricity or other energy services, and their ability to pay for these services (see IEA, UNEP and UNIDO, 2010). Questions of access and affordability apply between as well as within countries. For example, the World Energy Council has recently underlined the significance of *universal* access to affordable (and modern) energy services (World Energy Council, 2016). On a national level, questions of energy or fuel poverty – the inability to afford to keep one's home adequately heated – tend to focus on the cost of energy (Heffron and McCauley, 2017), rather than on the provision of services as such (see Chapter 3). Methods of tackling fuel poverty consequently focus on fiscal policies designed to help less-affluent citizens buy the energy they 'need'. According to recent research by Demski, the UK public think that affordability is more important than the other poles of the trilemma (Demski et al., 2017).

Low carbon supply

The third big challenge of energy trilemma is that of decarbonising energy supply. Following commitments to CO_2 emissions reduction targets, which began in the late 1980s and gained momentum through the 2000s, there is increasing pressure to reduce energy-related carbon emissions. Measures adopted in response include efforts to increase energy efficiency and the development of more renewable and low carbon forms of energy supply (World Energy Council, 2016). Lately the emphasis has been on decarbonising the electricity sector as electricity's role in transport and heating (and cooling) is expected to grow (ibid.).

Although it is widely agreed that all three aspects are important, they are linked to very different areas of policy responsibility including welfare, business and innovation, sustainability and technological efficiency. Not surprisingly, there is a tendency to treat each 'pole' as a distinct problem, framed with reference to different arguments, approaches, metrics and policy objectives. Rather than pointing to a single solution, responding to the trilemma as a whole depends on the mutual calibration of a mix of policies. In this role, the trilemma underscores the need to bring different forms of energy governance together in a single frame (Goldthau, 2011), but as Gunningham (2013) notes, studies and initiatives encompassing all three dimensions remain rare.

Balance or choice?

The energy trilemma is commonly depicted as an equilateral triangle in which relations between the three dimensions are indicated by lines or double headed arrows (Figure 10.1). This representation emphasises the ideal of achieving a perfect balance between the three priorities of security, lower carbon supply and affordability. From this point of view, effective energy policies are those which contribute to this ambition.

But as critics have observed, the more common pattern is for one or two of the three goals to be favoured over the other. For example, Heffron and McCauley (2017) argue that focusing on low cost and efficient solutions (affordability) has led to a continued reliance on fossil fuels in the short term, and at the expense of building low carbon energy infrastructure or developing a low carbon economy. To give another example, rather than being an outcome of 'balance', decisions to pass the costs of ensuring more secure, lower carbon supplies on to the consumer arguably represent a choice in favour of these two 'nodes' at the expense of consumer 'affordability'.

The relationship between choice and balance is also complicated. In so far as the energy trilemma is a 'true' trilemma in the philosophical sense, one or two of its aspects have to be compromised. Since policy makers have to make choices between the three policy goals involved, it follows that the trilemma can only ever be 'managed': it cannot be resolved (Gunningham, 2013).

Working with the trilemma

These observations are important in thinking about the trilemma's status within policy. If we accept that there is no 'solution', what part does the trilemma play in guiding policy action? One response is that it serves to remind policy makers (typically focusing on one or another of the three poles), that they should at least consider the consequences of their actions for the energy system as a whole. Another is that it provides a point of reference and a benchmark in terms of which policies can be evaluated. As mentioned above, the World Energy Council has come up with an aggregate national-level measure – the 'energy trilemma index' – with the aim of increasing the transparency of energy governance, and of identifying and benchmarking 'good' or more balanced responses. The index ranks countries in terms of the emphasis they place on energy security, affordability and sustainable provision, and provides a measure of how 'well' they handle these competing goals.

Since 2010, 125 countries have been given a 'balance score' that represents their performance in these terms (World Energy Council, 2017). The Energy Trilemma Index turns a very complicated policy concept – which already consists of competing policy ambitions – into a single measure.

This makes it possible to produce regional and country-specific profiles, and to enable comparison. In use, the index shows that countries often excel in one dimension of the energy trilemma but struggle to balance all three priorities. For instance, many of the countries performing well in the energy security dimension – such as Russia, US and Qatar – have extensive fossil fuels reserves, including coal, oil, and gas. By contrast, the countries that rank high in terms of environmental sustainability tend to be those that have a lot of geothermal or hydropower (both of which have low GHG emissions). Meanwhile, these same countries score less well in terms of 'security' because of what seems to be an over-reliance on a single energy source. In addition, and as Putra and Han observe, there are significant differences in how the trilemma is 'managed' in developed and developing countries, some of which are wary of risking economic growth for the sake of mitigating climate change (Putra and Han, 2014).

There are lots of questions one might ask about the basis and the purpose of this index. For instance, what are the units of analysis (in each area); what constitutes a 'good' balance; and what is the policy relevance or purpose of this measure? Questions about the scale at which 'the energy trilemma' arises are also important: is it something that affects households, regions, or only nation states. Equally and perhaps more importantly, trilemma-based discourses and policies do not explicitly engage with vital questions of demand. In the next section we draw attention to this missing part of the story.

The dilemma of the trilemma: where is demand?

As a concept, the energy trilemma is all about supply: is supply secure, is it affordable, and is it low carbon? In so far as policy goals are specified in these terms, they also focus on matters of supply. But what about the details of consumption and demand? How do changing patterns of energy demand figure in the trilemma, if at all?

The World Energy Council's energy trilemma index does take note of growth in energy consumption, but only on the basis that it presents challenges for security of supply. A change in energy consumption relative to growth in GDP is one of the measures that affects a country's rating in terms of energy security. As a result, some of the most significant fossil fuel economies, such as Saudi Arabia and the United Arab Emirates (UAE), are not ranked highly in the World Energy Trilemma Index (World Energy Council, 2016). Demand also figures in the guise of 'demand management'. For example, the World Energy Council claims that 'efforts to increase resource productivity and manage energy demand growth will be key in ensuring a balanced energy trilemma' (ibid., p. 4).

But what is meant by demand in this context? Often demand is equated with efficiency or associated with arguments about the scope for decoupling economic growth and CO_2 emissions. As such it is implicitly located within the 'lower carbon' pole of the trilemma. If demand is interpreted as the 'need' for fossil fuel, innovations in supply – like the introduction of more renewable energy – qualify as forms of 'demand reduction'. For example, Italy has recently set 'a target for renewable electricity generation – set at 10% by 2020 – to help counteract increasing energy demand and reduce GHG emissions' (World Energy Council, 2016).

More commonly, the level of demand – in the more fundamental sense of how much energy a society uses – is simply taken for granted. Chilvers et al.'s representation of future energy demand in the UK is typical:

> End-use energy demand is likely to remain roughly around its current level, although the energy transition out to the mid-21st century will require some switching towards greater electricity use, particularly for heating and transport. Consequently, achieving the UK CO_2 emissions reduction target will require a greater emphasis on systems for producing, delivering and using energy that is not only low carbon, but also secure and affordable for consumers both large and small.
>
> *(Chilvers et al., 2017, p. 441)*

Similarly, in a report produced by the World Energy Council, energy efficiency and so-called demand management strategies are deemed important in ensuring that countries continue to meet what are assumed to be unwavering and often growing needs (see Chapter 8).

> Energy efficiency and managing energy demand continue to be globally perceived as top action priorities with huge potential for improvement. . . . Policymakers must align the interests of asset owners, users and regulators, and continue to implement a combination of energy efficiency standards, performance ratings, labelling programmes and incentives.
>
> *(World Energy Council, 2016)*

As these examples show, discussions of the trilemma and how it might be managed do not engage with issues of demand head on. Instead, demand is typically equated with present levels of consumption, or with some very broad expectation of increasing future demand. This is consistent with a broader tendency to take the 'need' for energy as a given and to focus on methods of meeting demand more efficiently (Shove and Walker, 2014; Shove, 2018).

Although rarely discussed as a topic in its own right, levels and forms of energy demand are hugely significant for the trilemma, and for policy responses to it. One way of making this visible is to modify Figure 10.1, adding demand as a new dimension (Figure 10.2).

Turning the flat triangle of the energy trilemma into a pyramid immediately draws attention to issues of scale. By implication, the dimensions of the triangle (conventionally used to represent the trilemma) shrink or grow depending on the scale of demand. This is something of an over-simplification in that the three poles of security, affordability and low carbon supply are not fixed

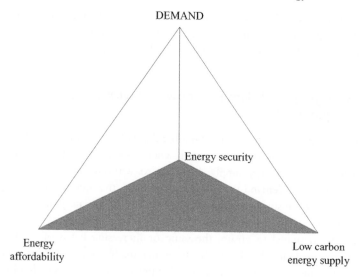

DEMAND

Energy security

Energy
affordability

Low carbon
energy supply

FIGURE 10.2 Demand and the energy trilemma.

together, nor are they relationally linked in quite this way. However, that does not undermine the importance of articulating the practical and political implications of the overall 'size' of the problem, or the value of showing that this is tied to the extent of demand.

Assumptions about the fixed or fluid nature of demand also matter, qualitatively, for the sorts of considerations that gather at each of the three poles. For example, methods of decarbonisation that make sense *assuming present levels of demand* are likely to be more costly and less affordable than the sorts of measures that might be required if demand, itself, fell by 50 per cent (Strbac et al., 2016). Rather than taking such possibilities into account, responses to the trilemma involve comparing and evaluating methods of delivering the same (energy) services more cheaply, more reliably or with 'cleaner' resources. Strategies of analysis and intervention based on the trilemma consequently disconnect the discussion of energy from a discussion of the social practices on which energy demand depends, and from questions about how, in what direction and at what rate energy demanding practices are changing.

This is a problem in that demand is not some pre-existing 'need' that systems of supply (energy and transport infrastructures and technologies) simply meet. Instead, social practices and related technologies, systems and infrastructures of provision constitute each other, and energy demand is both a consequence and a part of these dynamic arrangements. It follows that the scale and character of demand – and thus the size and character of the trilemma – is never static. This is important in that many of the challenges that the trilemma addresses are significantly affected and to an extent caused by increasing demand.

Challenging the trilemma – how and with what consequences?

The concept of the energy trilemma highlights linkages between the 'great challenges of energy policy' – namely energy security, energy affordability and low carbon energy supply. In doing so without explicit reference to the dynamics of demand it perpetuates the myth that questions about what energy is for are somehow outside the remit of legitimate policy response. The trilemma focuses on supply, and because of this, responses are at least implicitly designed to ensure the same or increasing levels of service. In overlooking demand, talk of the trilemma inadvertently sustains specific understandings of 'normal' and acceptable practice and favours technological responses and solutions that cater to these so-called needs. In this respect, trilemma-based discourses help perpetuate present (contemporary, Western) practices and in so doing help maintain (or extend) the 'size' of the triangle, and thus the scale of the problem.

This begs the question of whether it is possible to take demand seriously *within* and as part of the trilemma? At a minimum, such a move would involve reflecting on the unintended consequences of taking need for granted. More ambitiously, thinking along these lines might enable new and perhaps more challenging proposals for reducing demand by halting or reducing escalating expectations of (energy intensive) service provision. Such strategies would reconfigure all three poles of the trilemma and the ways in which these dimensions interact. Far from limiting the scope of intervention, such a response would highlight dynamic relations between issues of security, affordability and carbon *and* currently invisible processes involved in the constitution and transformation of demand.

Note

1 *Merriam-Webster Dictionary*, s.v. 'trilemma', https://www.merriam-webster.com/dictionary/trilemma [accessed 28 February 2019].

Further reading

Gunningham, N. (2013). Managing the energy trilemma: The case of Indonesia, *Energy Policy*, 54, pp. 184–193.

World Energy Council (2016). *World Energy Trilemma*. London: World Energy Council. Retrieved from www.worldenergy.org/wp-content/uploads/2016/05/World-Energy-Trilemma_full-report_2016_web.pdf [accessed 27 February 2018].

References

Blumsack, S. and Fernandez, A. (2012) Ready or not, here comes the smart grid! *Energy*, 37, pp. 61–68.

Boston, A. (2013). Delivering a secure electricity supply on a low carbon pathway, *Energy Policy*, 52, pp. 55–59.

Chilvers, J., Foxon, T. J., Galloway, S., Hammond, G. P., Infield, D., Leach, M. . . . and Thomson, M. (2017). Realising transition pathways for a more electric, low-carbon energy system in the United Kingdom: Challenges, insights and opportunities. *Proceedings of the Institution of Mechanical Engineers, Part A: Journal of Power and Energy*, 234(6), pp. 440–477.

Demski, C., Evensen, D., Pidgeon, N. and Spence, A. (2017). Public prioritisation of energy affordability in the UK. *Energy Policy*, 110, pp. 404–409.

Goldthau, A. (2011). Governing global energy: Existing approaches and discourses. *Current Opinion in Environmental Sustainability*, 3(4), pp. 213–217.

Gunningham, N. (2013). Managing the energy trilemma: The case of Indonesia. *Energy Policy*, 54, pp. 184–193.

Heffron, R. J. and McCauley, D. (2017). The concept of energy justice across the disciplines, *Energy Policy*, 105, pp. 658–667.

HM Government (2014). Innovation the only way to tackle energy 'trilemma'. Retrieved from www.gov.uk/government/news/innovation-the-only-way-to-tackle-energy-trilemma [accessed 7 August 2018].

HM Government (2017). Clean growth strategy. Retrieved from www.gov.uk/government/publications/clean-growth-strategy [accessed 7 August 2018].

IEA, UNEP and UNIDO (2010). Energy poverty: How to make modern energy access universal? Retrieved from www.globalbioenergy.org/uploads/media/1009_IEA_-_Energy_poverty.pdf [accessed 23 November 2018].

International Energy Agency (2014). Energy security. Retrieved from www.iea.org/topics/energysecurity [accessed 7 March 2018].

Putra, N.A. and Han, E. (eds). (2014). *Governments' responses to climate change: Selected examples from Asia Pacific*. London: Springer.

Sautter, J. A., Landis, J. and Dworkin, M.H. (2008). The energy trilemma in the Green Mountain State: An analysis of Vermont's energy challenges and policy options. *Vermont Journal of Environmental Law*, 10, p. 477.

Sharick, A. (2015) The UK's 'energy trilemma' downgrade should worry us, but there could be a fix. Retrieved from www.ukerc.ac.uk/network/network-news/the-uk-s-energy-trilemma-downgrade-should-worry-us-but-there-could-be-a-fix.html [accessed 6 December 2017].

Shove, E. (2018). What is wrong with energy efficiency? *Building Research and Information*, 46(7), pp. 779–789.

Shove, E. and Walker, G. (2014). What is energy for? Social practice and energy demand, *Theory, Culture & Society*, 31(5), pp. 41–58.

Strbac, G., Konstantelos, I. Annendi, M. Pollitt, M. and Green, R. (2016). *Delivering future proof energy infrastructure: Report for the National Infrastructure Commission.* Retrieved from www.gov.uk/government/uploads/system/uploads/attachment_data/file/507256/Future-proof_energy_infrastructure_Imp_Cam_Feb_2016.pdf [accessed 19 January 2018].

World Energy Council (2016). *World Energy Trilemma.* London: World Energy Council. Retrieved from www.worldenergy.org/wp-content/uploads/2016/05/World-Energy-Trilemma_full-report_2016_web.pdf [accessed 13 November 2017].

World Energy Council (2017). World Energy Trilemma. Retrieved from www.worldenergy.org/work-programme/strategic-insight/assessment-of-energy-climate-change-policy [accessed 13 November 2017].

11

FLEXIBILITY

Jacopo Torriti

Tigers can move at great speed, but when stalking their prey they slink low to the ground. When hunting, they are agile enough to spring into action in an instant and slow enough to wait for the right moment. Are energy consumers similarly flexible?

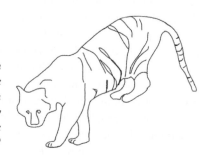

Introduction

Flexibility refers to the quality of bending easily without breaking and the ability to be easily modified. For a long time, energy demand was considered to be inflexible – non-negotiable and difficult to modify. Alongside the introduction of renewable energy technologies (which are widely seen as inflexible forms of generation), this view has recently changed: rather than being treated as something fixed, demand is now thought of as something that can and should be flexed, with the right kind of intervention. In exploring these themes, the chapter is in two parts. The first suggests that the recent history of energy provision supposes that demand is *not* flexible, while the second interrogates ongoing narratives according to which more and more flexibility is needed.

For the greatest part of the twentieth century and partly at the beginning of the twenty-first century, there has been a strong, supply-driven approach to energy provision. In this view, energy figures as a commodity (see Chapter 2), the need for which is primarily defined by the relationship between energy demand and economic growth (considered in more detail below). Following this logic, flexible supply was designed, planned and managed so as to meet variations in demand. However, much of what was known as flexible generation is disappearing as part of the phase-out of fossil fuel power plants. As a result, the *onus* of providing flexibility – in order to manage the system as a whole – is now falling on the 'demand side'.

Examples of the growing salience of 'flexibility' in the policy context in the UK include the creation of a flexibility team within Ofgem and, more recently the July 2017 Smart and Flexible Energy Plan by the Department for Business, Energy & Industrial Strategy (BEIS) and Ofgem (BEIS and Ofgem, 2017). The monetary value of flexibility (that is the value of the potential to shift loads in time) plays an important role in explaining the market interest in this area and the expanding emphasis on the feasibility of different forms of intervention. For instance, the value of the technical potential of the flexibility market was estimated at around £8 billion per year (National Infrastructure Commission, 2016). In this context, understanding how peaks are constituted, what demands are flexible, and what scope there is for effective load shifting becomes not only vital for the balancing of electricity demand and supply but also as a new form of market opportunity.

At the same time, flexibility is often interpreted very narrowly. For instance, it is commonly thought that flexibility in energy demand can only come about through deliberate technological intervention: for instance, in automated switching or via 'smart' devices (see Chapter 9). Techno-economic research tends to focus on loads that can be moved, mostly without 'disruption' to the operations of businesses and comfort of people (Silva et al., 2011). However, this does not take account of the different modes of flexibility that exist.

A broader definition of flexibility encompasses the capacity to use energy in different locations at different times of day or year (via storage or by changing the timing of activity, or by deferring or abandoning such action); to switch fuels; or, in the case of mobility, to re-arrange destinations and journeys in ways that momentarily reduce energy demand and/or congestion.

Research into flexibility consequently depends on a careful understanding of the timing of energy demand as something that is dynamic, negotiable, social, cultural, political and historical. As recent work in the DEMAND Centre at Lancaster University shows, the timing of energy demand is bound up with the temporal rhythm of society and with what people do.

This is obvious if we consider the substantial difference between residential electricity load curves for weekdays and weekends. During the same season the weather is often much the same during the weekend compared to the weekday. Building type, appliances, fuel substitution, price of energy and appliance control, and the moment of the day in which sunlight is present or absent may also be identical. The only substantial change between weekday and weekend is in terms of people's activities. Better understanding of these activities is critical not only in order to understand the timing of energy demand, but also in determining where and when the scope for flexibility really lies.

Whether this kind of knowledge will become central to the electricity sector or not depends on the way in which discourses and projects of 'flexibility' develop. To get a sense of the possibilities, we need to take a longer term view.

Inflexible energy demand: a short history

For some time, academic and lay conceptions of energy demand were structured around the idea that supply had to be modified in order to accommodate pre-determined levels of demand. This interpretation of inflexibility is itself an historical phenomenon, related to a specific moment in the development of networks of supply (extensive, and unending). This view is also rooted in certain disciplinary traditions, including physical–technical–economic modelling work, household economics and energy economics.

Most empirical studies of energy demand have tended to focus on one or several of the human 'factors' either causing or being consequences of household energy demand. The physical–technical–economic modelling work which has dominated energy analysis over the past 30 years or so places less emphasis on human occupants and more on building thermodynamics and technology efficiencies (Lutzenhiser, 1993). The technical and price data, along with the modelling techniques habitually used in energy demand research imply that the material and physical tend to prevail over the variable and correlational (Shipworth, 2013).

The main aspiration of existing studies has been to model and forecast household energy consumption rather than describe patterns of use or the basis of energy demand (Swan and Ugursal, 2009). Questions about how demand changes (e.g. throughout the day or over the year) are consequently neglected. Technical features like building design and appliance ownership bear some relation to volumes and aggregate levels of energy demand, but reveal little or nothing about intra-day variation or the timing of demand. In place of actual understanding of daily, weekly and seasonal fluctuations, many assume that patterns of consumption are repetitive and predictable (Suganthi and Samuel, 2011). The fixities at the heart of these assumptions about the repetitiveness and predictability of energy demand sustain the view that demand is essentially inflexible.

Other disciplines, including household economics, set much store by 'behavioural determinants' (which explain why appliances are used when

they are) when there are no obvious 'economic' explanations of how patterns of timing and energy demand come to be as they are. In the energy
economics literature, inflexibility has conventionally been associated with
inelasticity (see Chapter 6).

While flexibility refers to changes in demand due to a multiplicity of
factors, elasticity attributes changes in demand to peoples' responsiveness to
variations in price. Changes in price over short and medium periods yield
only minor changes in electricity demand (Borenstein, 2005; Filippini, 1995;
Silk and Joutz, 1997). According to this literature, two factors underpin the
inflexibility of residential electricity demand.

The first is the flat tariff to which most residential consumers are accus-
tomed. These tariffs obscure the true costs of electricity, as these change
over time. Other methods of charging are sensitive to time, but even then,
responses to peak period prices reveal price elasticities of 0.02 to 0.1 (Faruqui
and Sergici, 2009). Second, at least in the developed world, electricity prices
are usually 'too low' to make any impact on a large scale. Electricity prices are
typically low enough that electricity payments make up only a small portion
of the average household's budget, and consumers likely perceive electricity
as a necessity during peak times (Thorsnes, Williams and Lawson, 2012).

It is against this background that there is renewed interest in developing
and exploiting flexibility in demand.

Flexible energy demand: A long term future?

Accounts of why demand needs to be more flexible in the future relate to the
challenges of managing a higher proportion of more intermittent forms of
renewable energy supply (Barton et al., 2013). As already mentioned, there is
considerable interest in so-called demand side response involving storage and
back-up/standby assets, such as combined heat and power (CHP), which
can be dispatched at the request of power grid operators or of the plant
owner. The relatively recent literature on flexibility focuses on the need for
demand side response at different levels, i.e. national balancing, distribution
networks, etc. (Strbac, 2008; Bradley et al., 2013); the technical features of
flexible technologies (Siano, 2014; Hong et al., 2012) and alternative options
for the provision of flexibility (Poudineh and Jamasb, 2014).

Among all this, judgements about the extent and scale of flexibility in
the details and the timing of energy demand are central to projections and
narratives of a low carbon future in which intermittent renewables represent a significant share in the supply mix. The importance of demand-side
flexibility is captured in this extract of a report for the UK's National
Infrastructure Commission:

A lack of operational flexibility limits the system's ability to accommodate output from intermittent renewable technologies. This is a major concern since the decarbonisation effort will be based on replacing conventional plant with inflexible nuclear generation and intermittent wind and solar units. Naturally, more expensive low-carbon plant will be required to meet environmental targets under lack of operational flexibility if the system is not capable of fully absorbing renewable output.

(Strbac et al., 2016, p. 16)

As set out here, flexibility is needed as an alternative to expanding generation to provide extra 'capacity' (or more accurately, lack of demand) at times when demand and supply of electricity do not match.

The switch between interpretations of demand as fixed, on the one hand, and as an essentially flexible 'resource' on the other is as puzzling as it is striking. Is it really possible to manipulate demand in the ways that are implied and required by the shift towards more intermittent forms of supply? Was demand really inflexible in the era of 'predict and provide'? What happens if demand is not as flexible as it needs to be?

Challenging assumptions about flexibility

Energy researchers and policy makers are now assuming that aspects of energy demand that were previously taken for granted (and thought to be inflexible) can be made flexible. These shifting positions rests on sets of assumptions and historical conditions, five of which are highlighted below.

First, the idea that energy demand is inflexible (at any one moment) is underpinned by a raft of assumptions about economic necessity, needs and normality. Energy demand is expected to grow in conjunction with economic growth, and what is considered normal consumption is expected to change and increase over time as more people afford more energy-consuming services. Once specific interpretations of normality are established, it becomes impossible to reduce or modify the levels and timings of energy demand on which these 'needs' depend.

Second, energy infrastructures and systems of provision have traditionally been designed and constructed around production processes (typically industrial processes), and these have dictated the temporalities and geographies of the energy system. For instance, in developing countries, the periods of highest growth in energy supply have been historically coupled with periods of industrialisation. Most industrial processes were (and many still are) associated with large scale, reliable forms of energy supply. For instance, bulk chemical industries, and mining and refining have highly sequenced processes that depend on uninterrupted energy supply (Gutowski et al., 2009), as does semi-conductor manufacturing (Branham and Gutowski, 2010).

Third, it is increasingly recognised that energy systems are defined by different combinations of flexible and inflexible loads. For example, in their work on household energy use, Wood and Newborough (2003) distinguished between 'predictable', 'moderately predictable' and 'unpredictable' forms of consumption. Stokes, Rylatt and Lomas (2004) modelled domestic lighting with a stochastic approach and synthetically generated load profiles; and a study by Firth et al. (2008) analysed groups of electrical appliances (continuous and standby, cold appliances and active appliances) in terms of time of the day when they are likely to be switched on. In these types of analyses, loads, understood as the segments of demand associated with specific appliances and devices, have autonomous agency and either repeat themselves with the same temporalities or just happen in less predictable patterns.

Fourth, many believe that it is possible to generate and exploit flexibility in energy demand without interfering in the timing of consumers' energy-demanding activity. The hope is that this can be achieved via non-intrusive demand side response technologies that operate with little or no 'end user' engagement, or that form part of a contract with industrial and commercial customers who are financially rewarded either for being disrupted (i.e. turning down loads for pre-defined periods of time) or for providing off-grid generation (e.g. turning on standby generation where available).

Fifth, the energy system encompasses a huge range of flexible and inflexible technologies. The location of 'flexibility' is also variable in that the intended point of demand shifting can range from transmission and distribution lines through to individual household or commercial users. And at each point there are different technical constraints (Thrampoulidis et al., 2013). Stand-by generators are the main technology used in demand side response in the UK (Grünewald and Torriti, 2013). Alternative onsite generation technology consists of CHP boilers, wind turbines, and solar PV. Battery storage technology provides flexibility closer to home, particularly when it comes to managing site based generation and demand.

The five features described above illustrate the kinds of thinking involved in shifting understandings – from representations of demand as a fixed state of affairs, to new ideas about manipulating demand via price and technology intervention. They also illustrate the extent to which this thinking is rooted in a still limited understanding of the timing of energy demand.

Next steps in conceptualising flexibility

In everyday life, energy demands are endlessly variable, patterns of consumption are changing all the time, and doing so at different scales (Marsden and Docherty, 2013). This argues for conceptualising flexibility as an outcome of dynamic relations between technologies and practices, and of the intersection of multiple such arrangements.

This approach calls into question the idea that 'flexibility' is a property or quality of an individual. If people are treated as relevant 'units' of flexibility, then intervention should be measured in terms of how a variation in inputs (e.g. price or information made available through technology) affects individual demand. If practices are the unit of analysis, then changes in practices and how these are performed over time and space are the more relevant topics of analysis and investigation.

This approach also challenges ways of thinking about the 'drivers' of flexibility as if these were tied to the relation between price, or technology and energy demand. Assuming that social practices – such as weekend routines, meal times, or patterns of leisure – are appropriate 'sties' of enquiry, flexibility can and should be defined as an outcome of how these and other practices hang together in space and time.

Efforts to conceptualise flexibility are not frequent in the energy and social science literature, but there are some attempts in this direction. For example, Shick and Gad (2005) examine how the future 'flexible electricity consumer' is imagined in the Danish National Smart Grid Strategy and find that this vision relies on a techno-centric and rather 'inflexible' figuration of the consumer. However, there is clearly a need for more adventurous and also more conceptual research on the very idea of flexibility, and how it relates to the temporal organisation of energy demand. A distinctive and original approach along these lines depends on reconceptualising the time-related implications of flexible technologies, new pricing regimes and new forms of automation along with and as part of the transformation of social-temporal orders. Such an approach would enable researchers to engage with challenging questions about flexibility – in energy systems and in society, as they evolve together.

Further reading

Merton, R. C. (1998). Applications of option-pricing theory: twenty-five years later. *The American Economic Review*, 88(3), pp. 323–349.

Shick, L. and Gad, C. (2005). Flexible and inflexible energy engagements: A study of the Danish Smart Grid Strategy. *Energy Research & Social Science*, 9, pp. 51–59.

References

Barton, J., Huang, S., Infield, D., Leach, M., Ogunkunle, D., Torriti, J. and Thomson, M. (2013). The evolution of electricity demand and the role for demand side participation, in buildings and transport. *Energy Policy*, 52, pp. 85–102.

BEIS and Ofgem (2017). Upgrading our energy systems: Smart systems and flexibility plan. Retrieved from https://assets.publishing.service.gov.uk/government/uploads/system/uploads/attachment_data/file/633442/upgrading-our-energy-system-july-2017.pdf [Accessed 21 Nov 2018].

Borenstein, S. (2005). The long-run efficiency of real-time electricity pricing. *Energy Journal*, 26, pp. 93–116.

Bradley, P., Leach, M. and Torriti, J. (2013). A review of the costs and benefits of demand response for electricity in the UK. *Energy Policy*, 52, pp. 312–327.

Branham, M. S. and Gutowski, T. G. (2010). Deconstructing energy use in microelectronics manufacturing: An experimental case study of a MEMS fabrication facility. *Environmental Science & Technology*, 44(11), pp. 4295–4301.

Faruqui, A. and Sergici, S. (2009). *Household response to dynamic pricing of electricity – a survey of experimental evidence.* Retrieved from www.smartgrid.gov/files/ssrn_id1134132.pdf [accessed 15 February 2019].

Filippini, M. (1995). Swiss residential demand for electricity by time-of-use. *Resource and Energy Economics*, 17, pp. 281–290.

Firth, S., Lomas, K., Wright, A. and Wall, R. (2008). Identifying trends in the use of domestic appliances from household electricity consumption measurements. *Energy and Buildings*, 40, pp. 926–936.

Grünewald, P. and Torriti, J. (2013) Demand response from the non-domestic sector: Early UK experiences and future opportunities. *Energy Policy*, 61, pp. 423–429.

Gutowski, T. G., Branham, M. S., Dahmus, J. B., Jones, A. J., Thiriez, A. and Sekulic, D. P. (2009). Thermodynamic analysis of resources used in manufacturing processes. *Environmental Science & Technology*, 43(5), pp. 1584–1590.

Hong, J., Johnstone, C., Torriti, J. and Leach, M. (2012). Discrete demand side control performance under dynamic building simulation: A heat pump application. *Renewable Energy*, 39(1), pp. 85–95.

Lutzenhiser, L. (1993). Beyond price and attitude: Energy use as a social process. *Annual Review of Energy and the Environment*, 18, pp. 259–263.

Marsden, G. and Docherty, I. (2013). Insights on disruptions as opportunities for transport policy change. *Transportation Research Part A: Policy and Practice*, 51, pp. 46–55.

National Infrastructure Commission (2016). *Smart power.* London: National Infrastructure Commission. Retrieved from https://assets.publishing.service.gov.uk/government/uploads/system/uploads/attachment_data/file/505218/IC_Energy_Report_web.pdf [accessed 24 November 2018].

Poudineh, R. and Jamasb, T. (2014). Distributed generation, storage, demand response and energy efficiency as alternatives to grid capacity enhancement. *Energy Policy*, 67, pp. 222–231.

Shick, L. and Gad, C. (2005). Flexible and inflexible energy engagements: A study of the Danish Smart Grid Strategy. *Energy Research & Social Science*, 9, pp. 51–59.

Shipworth, D. (2013). The vernacular architecture of household energy models. *Perspectives on Science*, 21(2), pp. 250–266.

Siano, P. (2014). Demand response and smart grids – A survey. *Renewable and Sustainable Energy Reviews*, 30, pp. 461–478.

Silk, J. I. and Joutz, F. L. (1997). Short and long-run elasticities in US residential electricity demand: a co-integration approach. *Energy Economics*, 19, pp. 493–513.

Silva, V., Stanojevic, V., Aunedi, M., Pudjianto, D. and Strbac, G. (2011). Smart domestic appliances as enabling technology for demand-side integration: modelling, value and drivers. In T. Jamasb and M. G. Pollitt (eds), *The future of electricity demand: Customers, citizens and loads*. Cambridge: Cambridge University Press, pp. 185–211.

Stokes, M., Rylatt, M. and Lomas, K. (2004). A simple model of domestic lighting demand. *Energy and Buildings*, 36, pp. 103–116.

Strbac, G. (2008). Demand side management: Benefits and challenges. *Energy Policy*, 36(12), pp. 4419–4426.

Strbac, G. Konstantelos, I., Aunedi, M., Pollitt, M. and Green, R. (2016). *Delivering future-proof energy infrastructure*. Report for National Infrastructure Commission. Retrieved from www.nic.org.uk/wp-content/uploads/Delivering-future-proof-energy-infrastructure-Goran-Strbac-et-al.pdf [accessed 26 February 2019].

Suganthi, L. and Samuel, A. A. (2011). Energy models for demand forecasting – A review. *Renewable and Sustainable Energy Reviews*, 16, pp. 1223–1240.

Swan, L. G. and Ugursal, V. I. (2009). Modeling of end-use energy consumption in the residential sector: A review of modeling techniques. *Renewable and Sustainable Energy Reviews*, 13(8), pp. 1819–1835.

Thorsnes, P., Williams, J. and Lawson, R. (2012). Consumer responses to time varying prices for electricity. *Energy Policy*, 49, pp. 552–561.

Thrampoulidis, C., Bose, S. and Hassibi, B. (2013). Optimal placement of distributed energy storage in power networks. *IEEE Transactions on Automatic Control*, 61(2), 416–429.

Wood, G. and Newborough, M. (2003). Dynamic energy-consumption indicators for domestic appliances: Environment, behaviour and design. *Energy and Buildings*, 35, pp. 821–841.

12

NON-ENERGY POLICY

Sarah Royston and Jan Selby

Chameleons are known for changing colour, and for blending into the background. Many of the policies that matter for energy demand are also well camouflaged. This chapter investigates what are normally hidden forms of policy influence.

Introduction

What kind of policies affect energy demand? Most people would list policies about things like power stations and pylons, markets and meters, hobs and heaters. They might think of rules and regulations developed by energy departments, enshrined in energy strategies and implemented by energy managers. However, energy demand is not only affected by these explicitly energy-focused policies, but also by policies about trade, industry, transport, farming, health, education and many other issues.

Policies that are not explicitly about energy (which we call non-energy policies) matter for energy demand – probably much more than so-called energy policies. Non-energy policies play an important part in steering the things people do, like working, relaxing or moving around, and it is in doing these things that we use energy. However, the effects of non-energy policies are rarely recognised in research and government. The idea that 'energy policy is the only kind of policy that matters for energy demand' is widespread and deeply entrenched in policy and practice. Challenging this idea could bring about a step change in how energy and demand are understood, made visible and managed.

What is energy policy?

The assumption that energy policy is the only kind of policy that matters for energy demand is deeply embedded in the practice of many different organisations and institutions (whether it is made explicit or not). But what is meant by 'energy policy'? Energy policies are usually taken to be those dealing with the production, distribution and consumption of energy. For example, the remit of *Energy Policy* (an academic journal) encompasses issues of energy and environmental regulation, energy supply security, the quality and efficiency of energy services, the effectiveness of market-based approaches and/or governmental interventions, technological innovation and diffusion, and voluntary initiatives.[1]

On paper, definitions of energy policy tend to give equal weight to matters of supply and demand, but this is not always reflected in practice. For example, one textbook entitled *Understanding Energy and Energy Policy* (Braun and Glidden, 2014) has seven chapters on energy supply and none on energy demand. Similarly, policy makers remain mostly concerned with energy production, so much so that the demand side represents 'the Cinderella of energy policy, receiving scant policy attention and limited financial support when compared to energy supply' (Smith, 2009, pp. 64–65). Very recently, the UK government's Cost of Energy Review has been criticised as 'highly skewed towards considering supply-side issues and away from demand side policy' (End Use Energy Demand Centres, 2018).

Often 'energy policy' is in fact energy supply policy, and when energy policies do address demand, they tend to take a narrow view (see Chapter 2). Often, 'demand-side' policy really means *efficiency* policy; for example the UK Committee on Climate Change divides approaches to emissions reduction into two categories: 'using energy more efficiently' and 'switching to low-carbon fuels' (Committee on Climate Change, 2018a). In the recent past, efficiency policies in the UK have included obligations on energy companies to insulate homes (such as the Energy Company Obligation); standards and codes for buildings and appliances (such as the Code for Sustainable Homes); and financial incentives for energy improvements (such as the Green Deal, lower VAT on energy-saving materials and scrappage schemes). These do not tackle changing expectations, for example of comfort or home heating, at source.

To reiterate, energy policies are usually: (1) explicitly aimed at energy – as a topic in its own right; (2) largely focused on supply; and (3) if addressing the demand side, then mainly focusing on efficiency, often through technological improvements. Policies like these may have some impacts on energy consumption; for example, Rosenow and Galvin (2013) estimate that a loan scheme for home energy efficiency improvements in Germany achieved an average saving of 27 per cent of pre-refurbishment energy use. However, initiatives of this kind address only a tiny part of what matters for energy demand.

What other policies matter for energy demand?

To reduce energy demand, it is necessary to understand the full range of policies that have a bearing on how, where and when energy is used in society. Energy is not used for its own sake, but rather as part of what people do in daily life: for example, travelling, working or relaxing. These activities are affected by many forms of policy (by 'policies' we include standards, rules, regulations, and procedures for planning and oversight adopted by organisations ranging from international bodies to specific institutions like hospitals, councils and businesses), developed across many sectors.

A review by Cox, Royston and Selby (2016) found that energy demand (and supply) may be affected by policies on agriculture; trade; the economy; employment; welfare; health; education; defence; international development; the environment and devolution, among others. In some cases, the effect may be relatively direct and immediate; for example, if a local authority tries to improve road safety by providing additional street lighting, this will increase energy consumption as soon as the lights are switched on. Equally, if they decide to cut costs by reducing street lighting, there will be an immediate reduction in energy use. In other cases, the effect is long-term, indirect, delayed or occurs at a distance. For example, national and local land use policy decisions that support the development of out of town shopping centres have, over a period of time, had significant upward impacts on local transport-related energy demand (Banister, 1999).

Some 'non-energy' policies have a bearing on the timing of the activities that result in energy demand: when they happen, how frequently, for what duration and in what order. For example, Blue (2017) shows how targets and strategies in NHS hospitals (such as promoting a 'one-stop shop' approach to cancer diagnosis and treatment) have shifted the timings of treatments and appointments, with impacts on the resources used for patient care, including heated spaces and clinical equipment and on when related forms of energy demand occur. Since hospitals are major consumers of energy, these institutional rhythms have potentially large implications for the UK's overall energy demand.

Looking at the national level, the timing of peaks in electricity demand and in transport are related to long-term changes in employment including changes in working hours, male and female participation in the labour market and the trend towards working from home (Torriti, 2015; Torriti et al., 2015). Developments like these are linked to diverse policies, from how employment contracts are defined to how state welfare benefits incentivise people to move into work.

Other non-energy policies affect the spatial characteristics of activities in ways that matter for where energy demands arise, at what scale, and with what consequences for the mobility of people and goods. For example, education policies that promote free choice of schools have led to children travelling further to 'preferred' schools (He and Giuliano, 2017; Marshall et al., 2010). On a global scale, trade liberalisation policies favouring the movement of heavy industrial production to less developed countries have had huge consequences for where energy is consumed (Morgan, 2011). Such policies also increase the volume of goods transported around the world, with implications for shipping, aviation, road and rail transport. A large range of policies, from those on tourism through to defence, employment, international aid and development intentionally or unintentionally modulate flows of people, affecting where, when, how often, in what numbers, and by what means they travel around the world.

There are also wider-ranging policy agendas that affect energy demand, often doing so across many sites and sectors at once. In the UK, one of the most important is the commitment to economic growth, along with related strategies of commodification and marketisation. For example, the 'marketisation' agenda has been driving changes in UK higher education since 2000, including the introduction of student fees. This has led universities to invest in services and infrastructures – offering en suite accommodation and libraries that are open 24/7 to attract students, but that also have the effect of increasing energy demand (Royston, 2016).

Because policies like these interact with all kinds of social, technical and economic changes (which are themselves often linked to other policies) it is hard to identify and quantify their impact on energy demand. Where the impacts of non-energy policies can be traced, their effect – at many sites, sectors and scales – tends to be that of ratcheting up energy demand. Is this a pattern that can be reversed, and if so, how?

Mobilising non-energy policy to reduce demand

The effects of non-energy policies are both obvious – they are clearly significant for energy demand – but also invisible, and only rarely discussed

(Royston, Selby and Shove, 2018). For example, the UK's Committee on Climate Change states that demand-side measures are 'an essential part of the strategy to lower carbon emissions in the UK. . . . There will always be a demand for energy, but the way we use it, and the amount we use, needs to change' (Committee on Climate Change, 2018b). Having said that, the government projections on which models of future energy demand depend do not take account of the present and future role of non-energy policies in shaping demand.

The separation of energy from non-energy policy is reproduced and repeated across many organisational structures. For example, the energy or facilities managers who work in businesses, universities or hospitals rarely have the capacity or experience to intervene in what the institution *does* or in the forms of energy demand that follow. Equally, questions of energy seldom figure on the agendas of those in charge of 'mainstream' forms of planning and management. In such contexts, there is no opportunity to consider the consequences of non-energy policy (for instance, improving student experience, maximising grant income, promoting internationalisation and so on) or what these mean for the challenge of meeting institutional energy or carbon reduction targets.

The Committee on Climate Change has stated that 'emissions will not continue to fall without new and strengthened policies' (Committee on Climate Change, 2017, p. 8), but the question is whether and how policy makers (not just those dealing with energy) might come to terms with the challenge of energy demand reduction.

A first step in this direction is to recognise that energy demand and non-energy policies are linked, and to act on that insight. There are precedents for such an approach; for example, it is now widely recognised that non-health policies and practices across society affect patterns of wellbeing and disease (Egger and Swinburn, 1997). This recognition has had far reaching consequences for how causes, effects and problems are conceptualised, and for (some) of the policies developed in response. Optimising health now involves action to support physical safety at work (health and safety policies), to support people with physical and mental impairments (occupational health policies), to regulate food standards, and so on.

There is also scope to borrow concepts like that of the 'obesogenic environment' (Egger and Swinburn, 1997), which refers to the many arrangements that hold practices of eating and exercise in place (including the availability of green space, transport links, and the provision of different kinds of food) and develop similar approaches with respect to energy demand. This kind of thinking is already well established in the transport sector, where patterns of mobility are seen as outcomes of economic policy, urban planning and policies regarding education, leisure and employment, etc. Such approaches conceptualise mobility demand (and the energy used) not as a topic in its own right, but as something that is in large part 'derived' from other areas of daily life. The challenge is to bring the same logic to bear in relation to other forms of energy demand.

This calls for new kinds of research, as well as for more 'joined up' forms of policy intervention. For instance, we need to know more about precisely how non-energy policies and energy systems interact in order to identify and influence the diverse routes through which non-energy policies shape energy demand. To give a concrete example, such an exercise might involve unpacking the demand-related consequences of commitments like those of providing 'superfast broadband coverage to 95 per cent of UK premises by the end of 2017' (Priestley et al., 2017, p. 4). This would require thinking about the consequences of such a policy at different levels: how much energy do broadband technologies use; how much energy do related infrastructures require; how does broadband facilitate video streaming and other data-heavy practices; and how do such activities influence other seemingly unrelated practices and patterns of energy demand? This is just one case. Recognising the impact of non-energy policy on energy demand and actively working with non-energy policies to reduce demand opens up a new agenda for research and policy, and new opportunities for intervention and change.

Note

1 See *Energy Policy* website at www.journals.elsevier.com/energy-policy.

Further reading

Blue, S. (2017). Reducing demand for energy in hospitals: Opportunities for and limits to temporal coordination. In A. Hui, R. Day and G. Walker (eds), *Demanding energy: Space, time and change*. New York: Palgrave Macmillan, pp. 313–338.

Cox, E., Royston, S. and Selby, J. (2016). *The impacts of non-energy policies on energy systems: A scoping paper*. London: UKERC.

Royston, S., Selby, J. and Shove, E. (2018). Invisible energy policies: A new agenda for energy demand reduction. *Energy Policy*, 123, pp. 127–135.

References

Banister, D. (1999). Planning more to travel less: Land use and transport. *Town Planning Review*, 70(3), pp. 313–338.

Blue, S. (2017). Reducing demand for energy in hospitals: Opportunities for and limits to temporal coordination. In: A. Hui, R. Day and G. Walker (eds), *Demanding energy: Space, time and change*. New York: Palgrave Macmillan, pp. 313–338.

Braun, T. and Glidden, L. (2014). *Understanding energy and energy policy.* London: Zed Books.

Committee on Climate Change (2017). *Meeting carbon budgets: Closing the policy gap.* Report to Parliament. London: Committee on Climate Change.

Committee on Climate Change (2018a). What can be done. Retrieved from www. theccc.org.uk/tackling-climate-change/reducing-carbon-emissions/what-can-be-done [accessed 2 August 2018].

Committee on Climate Change (2018b). Using energy more efficiently. Retrieved from www.theccc.org.uk/tackling-climate-change/reducing-carbon-emissions/what-can-be-done/using-energy-more-efficiently [accessed 2 August 2018].

Cox, E., Royston, S. and Selby, J. (2016). *The impacts of non-energy policies on energy systems: A scoping paper.* London: UKERC.

Egger, G. and Swinburn, B. (1997). An 'ecological' approach to the obesity pandemic. *BMJ,* 315, pp. 477–480.

End Use Energy Demand Centres (2018). Energy efficiency: The missing piece in the energy cost review jigsaw – end use energy demand centres. Retrieved from www.eueduk.com/energy-efficiency [accessed 2 August 2018].

He, S. Y. and Giuliano, G. (2017). School choice: Understanding the trade-off between travel distance and school quality. *Transportation,* 44, pp. 1–24.

Marshall, J. D., Wilson, R. D., Meyer, K. L., Rajangam, S. K., McDonald, N. C. and Wilson, E. J. (2010). Vehicle emissions during children's school commuting: Impacts of education policy. *Environmental Science and Technology,* 44(5), pp. 1537–1543.

Morgan, N. (2011). *Carbon emission accounting: Balancing the books for the UK.* Energy Insight briefing paper. London: UKERC.

Priestley, S., Baker, C. and Adcock, A. (2017). *Superfast broadband coverage in the UK.* House of Commons briefing paper CBP06643. London: House of Commons Library.

Rosenow, J. and Galvin, R. (2013). Evaluating the evaluations: Evidence from energy efficiency programmes in Germany and the UK. *Energy and Buildings,* 62, pp. 450–458.

Royston, S. (2016). *Invisible energy policy in higher education.* Lancaster: DEMAND. Retrieved from www.demand.ac.uk/wp-content/uploads/2016/05/Royston-Invisible-energy-policy-in-HE.pdf [accessed 1 August 2018].

Royston, S., Selby, J. and Shove, E. (2018). Invisible energy policies: A new agenda for energy demand reduction. *Energy Policy,* 123, pp. 127–135.

Smith, A. (2009). Energy governance: the challenges of sustainability. In I. Scrase and G. MacKerron (eds), *Energy for the future: A new agenda.* Basingstoke: Palgrave Macmillan, pp. 54–76.

Torriti, J. (2015). *Peak energy demand and demand side response.* London: Routledge.

Torriti, J., Hanna, R., Anderson, B., Yeboah, G. and Druckman, A. (2015). Peak residential electricity demand and social practices: Deriving flexibility and greenhouse gas intensities from time use and locational data. *Indoor Built Environment,* 24(7), pp. 891–912.

13

POSTSCRIPT

Can energy researchers and policy makers change their spots?

Elizabeth Shove, Jenny Rinkinen and Jacopo Torriti

One of Rudyard Kipling's 'Just so' stories is about how the leopard got his spots. Can energy researchers and policy makers change their spots and work with different terms and concepts?

Introduction

Fables are short stories, typically with animals as characters, conveying a moral. When widely told and regularly repeated they become taken-for-granted truths – normative guides that steer action. In reviewing some of the fables that populate the energy sector, the essays in this collection interrogate recurrent themes and established narratives. They do so at a moment when the field is in flux. The scale of what is involved in meeting carbon reduction targets is gradually becoming apparent. The supply mix is changing with the introduction of renewable sources of energy and with the electrification of services such as heating and transport. Smart meters are generating more data than has ever been available, and there is renewed interest in issues of flexibility and timing as these affect the balancing of supply and demand. Questions about when energy is used raise others about what it is used for, and about how energy-demanding practices vary and change.

Meanwhile, dominant metaphors and metrics remain rooted in an era of seemingly endless power, and in ideologies of individual consumer choice, market competition and resource economics.

One reason for writing this book is to address this tension: to revisit the terms in which energy is known and understood and to extend and refresh the range of concepts and approaches in play. In questioning aspects of received wisdom and introducing new ideas, we are interested in reconfiguring problems and identifying new opportunities for significant reductions in carbon emissions. We have suggestions to make, but these do not add up to a list of immediate prescriptions for energy policy, not as it is currently formulated. Instead, and as detailed below, our contribution is to reveal the blind spots associated with current approaches, and to show how these might be overcome. In the next section we take stock of three recurrent themes – the representation of energy, the 'making' of energy demand, and the relations between technology and practice. This sets the scene for the final part of the book in which we reflect on the work that energy fables do in policy and practice.

In general terms, all fables simplify, and all forms of simplification have strings attached, in the sense that they reproduce certain perspectives and positions. The dominant discourses we consider in this collection are no exception. Since languages and concepts are 'performative' – meaning that they have effect on what people do – it should come as no surprise to find that the terms we discuss are enmeshed in a wider landscape of policy and practice. In this context, it would be naïve to expect researchers and policy makers to simply 'change their spots' or to adopt significantly new approaches any time soon.

It is equally naïve to take fables entirely at face value, or to suppose that research and policy making are of a piece. Behind the scenes, calculations of efficiency and of rebound reproduce a myriad of social and political judgements about wellbeing, service and normality. Similarly – but again invisibly – discussions of the trilemma embody and also disguise profoundly important assumptions about the nature of energy demand. Our further ambition, in this postscript and in the book as a whole, is to reveal these normally hidden aspects, and to draw attention to the ambiguous and sometimes contested settings in which they have effect.

On both counts – that is, in revisiting energy-related vocabularies and in revealing taken for granted truths – our aim is to prompt people involved in the energy sector to think again about the terms and concepts they use, the assumptions on which these depend and the actions that follow. In working through this book readers will come to their own conclusions about the issues we raise, and about the potential for developing a new generation of ideas and terms, consistent with the challenges that lie ahead. In bringing this book to a close we offer our own thoughts on the themes we have raised, and the implications these have for a new 'landscape' of energy-related research and policy.

Themes and implications

Each of the fables we have discussed reproduce different aspects of received wisdom. We now draw some of these threads together. Rather than itemising the policy implications of our insights and critiques, one at a time, we revisit three cross-cutting themes, starting with the terms in which energy is represented, measured, managed and known.

How energy is represented and known

Various fables (elasticity, efficiency, rebound, the trilemma, low hanging fruit) treat energy as a singular resource and as something that can be quantified in terms like million tonnes of oil equivalent or kilowatt hours. It is easy to forget how important this is for business-as-usual in the energy sector. Being able to calculate and compare units of energy is a necessary precursor for organising markets across different fuels and different types of energy generation, for discussions of efficiency (more service for the same units of energy), and for estimating price elasticity (willingness to pay for units of energy). If they are to fulfil this role, useful representations depend on abstracting energy from specific and situated practices of production, distribution and use. This is a necessarily selective process. For example, measuring the number of kilowatt hours associated with cooking, washing or cooling flattens out potentially important questions about how and when these different activities are done, how they interact and how they change over time. Put simply, different metrics enable certain discussions and lines of enquiry (e.g. appliance efficiency) but make others more difficult (e.g. shifting routines, new practices). They also influence the types of explanation that follow. Because current methods and measures treat energy as a generic resource they tend to obscure the *different* dynamics that underpin such diverse 'end uses' as heating, computing or laundering. Instead, fluctuations in demand are attributed to generic 'laws of the market', or to price and consumer choice rather than to infrastructural arrangements or institutional and collective interpretations of normality and need.

What are the policy implications of these observations? Are there ways of representing energy demand not in the abstract, but as part of what people do, and if so what new agendas might such approaches open up?

There are precedents that illustrate the value of working with different terms and descriptions. For example, the 'societal synchronisation index' (Torriti et al., 2015) captures the extent to which similar energy-intensive practices are done at the same time in any one society. In bringing issues of timing and demand together, it provides a distinctive method of characterising and measuring flexibility and the potential to reduce peak load.

Other techniques have been used to show how energy services change over time. While the rise in indoor temperatures in UK homes is well documented (Palmer and Cooper, 2013), more subtle social historical research demonstrates that meanings of comfort (and of activity, clothing, seasonality) have shifted alongside and as part of these developments (Shove, 2003; Humphreys, Nicol and Roaf, 2011; Kuijer and Watson, 2017; Trentmann and Carlsson-Hyslop, 2017). There is no standardised measure of how expectations and conventions evolve and no yardstick with which to evaluate the part that policy makers play in these processes. However, studies like these have the very important effect of revealing the extent to which energy demands and forms of energy supply are woven into the details of everyday life, into decisions about fuels and infrastructural investment and into the policies and strategies of local and central governments alike.

There is clearly more to be done but these few examples suggest that it *is* possible to describe and represent normally 'hidden' aspects of the energy system. Looking ahead, methodological innovations of this kind are vital if matters of timing, synchronisation and demand are to become established and legitimate topics of negotiation and debate.

How energy demand is made and not simply met

Many energy-related fables (energy services, energy efficiency, keeping the lights on, the energy trilemma) take demand for granted. For example, discussions about the 'trilemma' are in essence discussions about how to handle tensions between energy security, affordability and carbon – but not about how much energy is required or for what purpose. In some cases, questions of demand are marginalised because they are thought to be matters of consumer choice, and as such, 'off limits' for policy makers and researchers alike. One consequence is that responses and strategies are developed and evaluated as if demand was simply 'there'.

We argue that this is misleading and counterproductive on two counts. First, and most obviously, present ways of life are unlikely to remain the same forever, meaning that future patterns of demand are unlikely to match those with which we are familiar with today. Second, and more important, how demands develop is, in no small measure, influenced by past and present policies, infrastructures and technologies. As is often said in transport studies, demand is 'derived' from other activities that are shaped by non-energy policies, including policies on housing, education, health and so forth. From this point of view, policy makers and researchers are *always* involved in constituting future demand, whether they are aware of it or not.

Seeing and perhaps redirecting these forms of policy influence calls for a more transparent approach to dealing with demand and for articulating assumptions that are 'baked' into energy modelling, into related policies, and into current programmes of technological research and development. As already explained, the unquestioning commitment to 'keeping the lights on' stands for a broader, also unquestioning commitment to meeting present and future demands, whatever those might be.

In short, there is scope for much more adventurous and much more challenging policy making informed by debates about what substantially less 'demanding' ways of life might be like and about the kinds of material and institutional arrangements on which these might depend. Building on these ideas, there is a related need to link energy and non-energy policies and to actively use the latter to reduce demand and/or influence the timing and location of energy consumption. This is not a novel idea, but if such suggestions are to take hold it will be vital to recognise that demand is not fixed: it is open to negotiation, and is continually on the move. Developing this agenda depends on thinking again about how technologies and practices shape each other.

How technologies and practices shape each other

It is obvious that energy provision, distribution and use are mediated by technology. What is disturbing is that discussions of smart technologies, flexibility, energy demand, energy services, and energy efficiency generally take the 'functions' of appliances and infrastructures for granted. For example, more efficient heating or lighting systems are expected to take the place of those they replace, and to do so without disrupting established conventions of comfort. This is consistent with the tendency to distinguish between 'technologies' and 'users' and to focus on methods of persuading people to 'take up' or adopt innovative solutions.

It is now widely recognised that energy systems are 'sociotechnical'. However, there are different interpretations of what this means. Some authors use the term when describing the social and institutional shaping of transition pathways (Geels and Schot, 2007) or when itemising social factors that govern the uptake of lower carbon technologies. Others refer 'sociotechnical' systems in order to make the simple point that people are involved – for instance, in using and adapting buildings and appliances (Chiu et al., 2014). Although all take note of social processes, analyses like these generally overlook the extent to which technologies are implicated in the conduct of daily life, and in defining the activities of which they are themselves an integral part.

The contributors to this collection contend that technologies are social in a much more profound sense. In becoming 'configurations that work' (Rip and Kemp, 1998) appliances and infrastructures are integrated into, and constitutive of the social practices of which societies are made. This way of thinking has important policy implications. For example, instead of figuring out how to seed a low carbon transition by promoting a suite of technological solutions, the more subtle challenge is that of understanding how material arrangements enable and constrain different complexes of social practice and what these mean for the resource and energy demands that follow. Some might worry that this is a step too far for energy policy makers, especially given the 'siloed' demarcation of expertise, and of related roles, responsibilities and remits. However, it is important to recognise that whether they are aware of it or not governments and businesses are *inescapably* involved in defining what count as 'normal' standards and ways of living. Acknowledging that policy is part of and not outside these dynamics is a precondition for identifying, for better understanding and for intervening in the trajectories and processes involved.

Moving in this direction depends on moving away from generic models of energy demand, and towards more historically and culturally precise accounts of how specific energy-demanding practices come to be as they are, and how they are in any case changing. Different languages are needed to represent these processes, and to describe how energy-demanding practices emerge and disappear, and how they circulate and combine (Hui, Schatzki and Shove, 2017).

The terms and phrases discussed in this book reproduce 'received wisdom' in the energy field, and in so doing work together to sustain and reproduce a body of ideas about what is normal and about what drives change. The result is a dominant paradigm that is increasingly out of step with the challenges now facing the energy sector. On the face of it, this situation calls for new approaches and ideas – perhaps along the lines suggested above. However, this supposes that policy makers and researchers can simply 'change their spots'.

Can energy researchers and policy makers change their spots?

Schools of thought, and the fables they give rise to, do not emerge by chance. Instead, 'spots' – by which we mean conceptual frameworks, methodologies and styles of analysis and problem framing – are anchored in professional and institutional identities and in disciplinary traditions. In practice, this means that the theoretical foundations of energy research and policy are unlikely to be overhauled any time soon. On the other hand, policy and research are not always of a piece. As a result, competing and sometimes contradictory positions *already* exist.

For policies to have effect, they have to enter and become part of the flux of what different people do. In other words, they have to become embedded in settings that have quite specific histories and contexts. Making and implementing policy is thus a pragmatic business, and one that happens in real time. Ironically, these complexities are ironed out of policy-oriented research much of which takes place within what Lutzenhiser describes as a 'looking glass world' – a parallel universe populated by generic methods of modelling and economic analysis, largely devoid of history, and stripped of 'extraneous' and potentially complicating factors (Lutzenhiser, 2014).

It is in these decidedly ambivalent settings that the fables we have discussed are reproduced, and in which they have effect. In many situations, concepts of efficiency, or elasticity along with ideas about low hanging fruit and the promises of smart technologies inform responses to seemingly real problems. However, there are other instances in which these terms and concepts are known to be useful (perhaps necessary) fictions: they help organise research and policy, and they are part of constituting and not just responding to challenges within the energy sector. If we were to simply offer alternative and potentially better or more convincing narratives we would be in danger of taking existing discourses and related method and problems at face value. To avoid this trap, we need to consider the work that fables do in organising and simplifying policy contexts that are in fact never that simple. This approach puts the positions and arguments we have discussed in a different light.

The language of energy efficiency illustrates this complexity. On the surface, proponents of energy efficiency are devoted to the task of delivering the same or more service but with less energy. At first sight, this means that the pursuit of efficiency has little or nothing to do with potentially contentious questions about how much energy is needed or what counts as sufficient. But if we look behind the scenes, it becomes obvious that efforts to promote efficiency have the necessary and unavoidable consequence of promoting some, and not other interpretations of sufficient, appropriate and acceptable meanings of service. Like it or not, efficiency measures carry with them an invisible baggage of normative commitment. The same applies to seemingly neutral and seemingly self-evident strategies like those of 'picking the low hanging fruit'.

On closer inspection, what appear to be purely economic logics of payback and return on investment prove to be infused with social and cultural assumptions. The details vary from case to case, but across the board established methods and metrics obscure, but also depend on a raft of tacit understanding and convention. For example, guidelines for energy efficient building reproduce understandings of normal indoor temperatures. Similarly, programmes of energy labelling in the office sector reproduce judgements about normal power loads, and so on. Exactly the same applies to discourses of energy security and to concerns about the energy trilemma, all of which rest on a platform of weakly articulated but pervasive assumptions about societal needs.

Ironically, and in all these cases, dominant discourses and fables that appear to marginalise questions of demand have the invisible but powerful effect of reproducing very specific interpretations of what energy is for and how much is needed. In other words many of the concerns addressed in our critiques, and many of the seemingly 'missing' debates and topics are not literally absent but are instead camouflaged by the terms in which the field is organised.

Which issues are visible and which are not, which topics are out in the open and which remain in the shadows? The patterns we have described are not random, nor can they be transformed by act of will. As we have seen, the phrases and refrains of energy research and policy reflect and reproduce boundaries and distinctions that are firmly grounded and historically situated. This landscape matters, but habits of thought do not stay still. The patterning of debates and the social and institutional structures to which they connect continue to evolve. This means that although individual energy researchers and policy makers are unlikely to change their spots – to cast familiar terms aside, or to make explicit that which is usually invisible – dominant discourses are always on the move, being propelled and sometimes transformed by discussions and reactions like those we hope this book provokes.

References

Chiu, L. F., Lowe, R. Raslan, R. Altamirano-Medina, H. and Wingfield, J. (2014). A socio-technical approach to post-occupancy evaluation: Interactive adaptability in domestic retrofit. *Building Research & Information*, 42(5), pp. 574–590.

Geels, F. W. and Schot, J. (2007). Typology of sociotechnical transition pathways. *Research Policy*, 36, pp. 399–417.

Hui, A., Schatzki, T. and Shove, E. (2017). *The nexus of practices: Connections, constellations, practitioners*. London: Taylor & Francis Group.

Humphreys, M., Nicol, F. and Roaf, S. (2011). *Keeping warm in a cooler house*. Technical paper no. 14. Edinburgh: Historic Scotland.

Kuijer, L. and Watson, M. (2017). 'That's when we started using the living room': Lessons from a local history of domestic heating in the United Kingdom. *Energy Research & Social Science*, 28, pp. 77–85.

Lutzenhiser, L. (2014). Through the energy efficiency looking glass. *Energy Research & Social Science*, 1, pp. 141–151.

Palmer, J. and Cooper, I. (2013). *United Kingdom housing energy fact file*. London: HMSO.

Rip, A. and Kemp, R. (1998). Technological change. In S. Rayner and E. L. Malone (eds), *Human choice and climate change, vol. II: Resources and technology*. Columbus, OH: Battelle Press, pp. 327–399.

Shove, E. (2003). *Comfort, cleanliness and convenience: The social organization of normality*. Oxford: Berg.

Torriti, J., Hanna, R. Anderson, B. Yeboah G. and Druckman, A. (2015). Peak residential electricity demand and social practices: Deriving flexibility and greenhouse gas intensities from time use and locational data. *Indoor and Built Environment*, 24(7), pp. 891–912.

Trentmann, F. and Carlsson-Hyslop, A. (2017). The evolution of energy demand in Britain: Politics, daily life and public housing 1920s–1970s. *The Historical Journal*, 61(3), pp. 807–839.

INDEX

Page numbers in *italics* refer to figures.